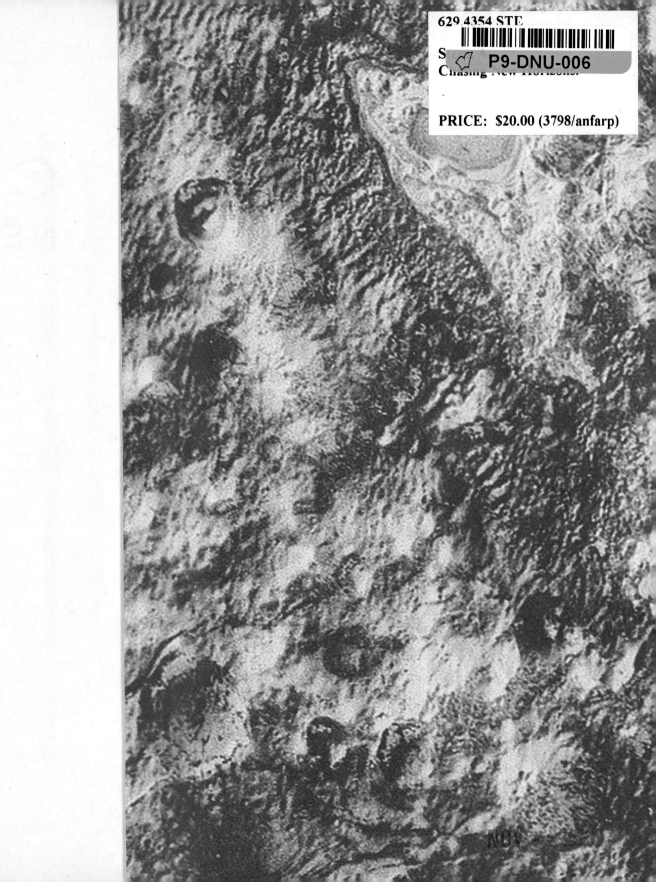

CHASING
NEW HORIZONS

ALSO BY ALAN STERN

Worlds Beyond: The Thrill of Planetary Exploration as Told by Leading Experts (editor)

Our Universe: The Thrill of Extragalactic Exploration (editor)

Our Worlds: The Magnetism and Thrill of Planetary Exploration (editor)

Pluto and Charon: Ice Worlds on the Ragged Edge of the Solar System (coauthored with Jacqueline Mitton)

The U.S. Space Program After Challenger: Where Are We Going?

ALSO BY DAVID GRINSPOON

Earth in Human Hands: Shaping Our Planet's Future

Lonely Planets: The Natural Philosophy of Alien Life

Venus Revealed: A New Look Below the Clouds of Our Mysterious Twin Planet

The Planet Venus (coauthored with Mikhail Marov)

CHASING
NEW HORIZONS

INSIDE THE EPIC FIRST
MISSION TO PLUTO

ALAN STERN AND
DAVID GRINSPOON

PICADOR

NEW YORK

picadorusa.com • instagram.com/picador
twitter.com/picadorusa • facebook.com/picadorusa

Picador® is a U.S. registered trademark and is used by Macmillan Publishing Group, LLC, under license from Pan Books Limited.

For book club information, please visit facebook.com/picadorbookclub or email marketing@picadorusa.com.

Endpaper art courtesy of NASA

Designed by Jonathan Bennett

Library of Congress Cataloging-in-Publication Data

Names: Stern, Alan, 1957– author. | Grinspoon, David Harry, author.
Title: Chasing New Horizons : inside the epic first mission tc Pluto / Alan Stern and David Grinspoon.
Description: New York : Picador, [2018] | Includes index.
Identifiers: LCCN 2017060114| ISBN 9781250098962 (hardcover) | ISBN 9781250098986 (ebook)
Subjects: LCSH: Space flight to Pluto. | New Horizons (Spacecraft) | Pluto probes. | Pluto (Dwarf planet)—Exploration.
Classification: LCC TL799.P59 S74 2018 | DDC 629.43/54922—dc23
LC record available at https://lccn.loc.gov/2017060114

Our books may be purchased in bulk for promotional, educational, or business use. Please contact your local bookseller or the Macmillan Corporate and Premium Sales Department at 1-800-221-7945, extension 5442, or by email at MacmillanSpecialMarkets@macmillan.com.

First Edition: May 2018

10 9 8 7 6 5 4 3 2 1

This book is dedicated to all the incredible men and women whose work contributed to New Horizons, to their families for supporting their dedication to its success, and to all the others who, along the way, helped make the exploration of Pluto possible.

CONTENTS

PREFACE

INSIDE THE FARTHEST
EXPLORATION IN HISTORY

In January of 2006, a tiny one-thousand-pound spacecraft, mounted on top of a powerful 224-foot-tall rocket, blasted off from Cape Canaveral, Florida. Thus began the longest and farthest journey of exploration our species has ever attempted—a journey to explore Pluto, the last of the unvisited planets known at the dawn of the Space Age. That spaceship, aptly named New Horizons, carried the hopes and dreams of a team of scientists and engineers who had poured much of their lives into what had—at many times—seemed an improbable quest.

Some sixty years ago, we humans began reaching across space—the final frontier—to explore other worlds. Before that, such explorations could only take place in works of fiction. But in this new era, we—sentient beings of the Sun's third planet—have begun to send humans and robotic ships across the vastness of space to explore other worlds. The period of time we live in will forever be known as the era when humans emerged from the cradle that is our planet, to become a space-faring species.

In the 1960s and 1970s, NASA's Mariner spacecraft made humankind's first successful journeys—to the closer planets—Venus, Mars, and

Mercury, and humans first walked on the Moon. Also in the 1970s, NASA spacecraft called Pioneer were the first to reach Jupiter and Saturn, much farther away than the inner planets. Following that came NASA's Voyager project, originally cast as the "grand tour" mission that would visit all five then known outermost planets, from Jupiter to Pluto. But in the end, Voyager did explore Jupiter, Saturn, Uranus, and Neptune but did not travel to Pluto. And so, as the 1980s closed, all the then known planets but one had been visited by spacecraft. As a result, Pluto, that lone, unexplored planet, became something more: it became to some a symbol, an open challenge, and a dare.

NASA's New Horizons mission to Pluto, which resulted and which we chronicle here, was a logical continuation of all those previous first journeys of planetary exploration. Yet New Horizons was also, and in many ways, like nothing attempted ever before. Many doubted that it could be approved, and even more doubted there was enough money or time to build it, or that it would ever succeed. But as we describe in this book, a dedicated and persistent group of scientists and engineers defied expectations and over twenty-six years, made an almost impossible dream of exploration come true, in 2015.

Our goal in this book is to give you a sense of what it took to get this landmark space mission conceived, approved, funded, built, launched, and flown successfully to its oh-so-far-away target. There are many aspects of this story that are emblematic of modern space exploration. Yet there were also incidents and episodes completely unique to the story of New Horizons: unforeseen hazards, threats, misdeeds, and misfortunes that had to be overcome, and there were many fortuitous moments where luck and good fortune played a key role and without which the quest would have never succeeded.

We, this book's authors, are two scientists whose involvement in New Horizons has been very different—one central, one peripheral. But we are joined in our shared excitement over both the exploration of faraway worlds and our wish to share the very special, fascinating, and largely untold story of New Horizons and how distant Pluto actually came to be explored.

Alan Stern's involvement is at the core of this story. Although literally thousands of people were involved in New Horizons, Alan was the project's leader from its very start. In contrast, David Grinspoon played only a tangential role in this story. Like Alan, David is a planetary scientist but he is also a writer by trade. For decades David has been a close friend and colleague of both Alan's and many other key participants in this story, and David was present at many of the pivotal moments in the saga. For example, David served on NASA's all-important Solar System Exploration Subcommittee in the 1990s and early 2000s, where, as you'll read, some of the crucial decisions were made that gave birth to New Horizons. And David was there at the raucous "win party" on Bourbon Street in New Orleans in 2001, when New Horizons had just been selected by NASA over proposals from fierce competitors. David was also at Cape Canaveral for the earsplitting, soaring launch to Pluto in 2006, and he helped the team devise strategies for public outreach surrounding the 2015 flyby of Pluto. When New Horizons explored Pluto, David worked with the science team as a press liaison to the media. Though many of David's impressions and descriptions here are firsthand, he is not often a named character in this book. Rather, his voice here represents the book's storyteller.

The two of us met twenty-five years ago, just after this story begins, and we've marveled at the series of unlikely events that has unfolded since that time, as we traveled through our lives, and as New Horizons fought for approval, was built, and traveled across our solar system.

In what follows, we've attempted to meld our voices to provide a combined and intimate perspective on the historic journey to conceive, create, and culminate in the exploration of Pluto—the capstone journey in the first reconnaissance of the planets of our solar system.

The core material of this book came out of a long series of phone conversations the two of us had, every Saturday morning for a year and a half, in which we retraced the long adventure of New Horizons as Alan told David his memories of the project and all its precursors and phases. Out of the transcripts of these chats between us,

David wrote the first drafts of most of the chapters, which we then both edited and rewrote many times, passing drafts back and forth and winnowing the narrative.

The result: this book synthesizes our two views of this amazing tale, supplemented by the voices of numerous other key players in the story. But this book is largely the story as seen through its leader Alan's eyes, as told to David.

Writing this book together presented some challenges. For example, how would we refer to Alan? We couldn't, except in quoted passages, use the first person (as in "I couldn't believe what I was hearing!") since David is a coauthor. Although it seemed a little strange to use the third person (as in "Alan couldn't believe what he was hearing!") in a book with Alan as coauthor, yet, for stylistic purposes, we did elect to use the third person to refer to Alan. Alan's voice, like the voices of others, appears in first-person quotes, often set off from the main text. Many of these quotes come from the transcripts of our Saturday conversations with which we began this project.

Modern planetary exploration is a complex effort that cannot succeed without the work of a great many people. Some of the participants in New Horizons spent decades of their lives dreaming, scheming, planning, building, and flying the one and only spacecraft ever sent to Pluto. So we want to acknowledge that many more individuals contributed to the exploration of Pluto than can be named in this book. That's regrettable to both of us, but in the interest of telling a manageable story, we sadly had to leave out many who made contributions. We thank our editors for improving our story by insisting it remain short enough to be read.

Nothing quite like New Horizons has occurred in a generation— the raw exploration of new worlds. And nothing quite like it is currently planned to happen, ever again.

In what follows, we share what it was like to be involved in New Horizons, one of the best-known projects in the history of space exploration. The effort to explore Pluto was an unlikely and sometimes harrowing story, with so many unexpected twists and turns, seeming dead ends, and narrow escapes that it hardly seems possible that it actually succeeded—but it did.

Come with us now and learn how it happened, and experience how it felt to be inside it all.

—ALAN STERN, Boulder, CO
DAVID GRINSPOON, Washington, DC
January 2018

CHASING
NEW HORIZONS

INTRODUCTION

OUT OF LOCK

On the Saturday afternoon of July 4, 2015, NASA's New Horizons Pluto mission leader Alan Stern was in his office near the project Mission Control Center, working, when his cell phone rang. He was aware of the Independence Day holiday but was much more focused on the fact that the date was "Pluto flyby minus ten days." New Horizons, the spacecraft mission that had been the central focus of his career for fourteen years, was now just ten days from its targeted encounter with the most distant planet ever explored.

Immersed in work that afternoon, Alan was busy preparing for the flyby. He was used to operating on little sleep during this final approach phase of the mission, but that day he'd gotten up in the middle of the night and gone into their Mission Operations Center (MOC) for the upload of the crucial, massive set of computer instructions to guide the spacecraft through its upcoming close flyby. That "command load" represented nearly a decade of work, and that morning it had been sent by radio transmission hurtling at the speed of light to reach New Horizons, then on its approach to Pluto.

Glancing at his ringing phone, Alan was surprised to see the caller was Glen Fountain, the longtime project manager of New Horizons. He felt a chill because he knew that Glen was taking time off for the holiday, at his nearby home, before the final, all-out intensity of the upcoming flyby. Why would Glen be calling now?

Alan picked up the phone. "Glen, what's up?"

"We've lost contact with the spacecraft."

Alan replied, "I'll meet you in the MOC; see you in five minutes."

Alan hung up his phone and sat down at his desk for a few seconds, stunned, shaking his head in disbelief. Unintentional loss of contact with Earth should never happen to any spacecraft. It had never before happened to New Horizons over the entire nine-year flight from Earth to Pluto. How could this be happening now, just ten days out from Pluto?

He grabbed his things, poked his head into a meeting down the hall where he was supposed to be heading next, and said, "We've lost contact with the spacecraft." Then his colleagues looked at him, dumb-founded. "I'm headed to the MOC and don't know when I'll be back. It probably won't be today." He walked out to his car into the Maryland summer heat, and drove the half-mile across the campus of the Applied Physics Laboratory in Laurel, Maryland, where New Horizons was operated.

That drive was probably the longest few minutes in Alan's life. He had high confidence in his team handling emergencies: they had re-hearsed so many contingency scenarios, if any team could handle this, it would be New Horizons. But still, he couldn't prevent his mind from picturing the worst.

Specifically, he couldn't help but think of NASA's ill-fated Mars Observer. That spacecraft, launched in 1992, also went silent, just three days before reaching Mars. All attempts to reestablish communication were unsuccessful. NASA later determined that Mars Observer had experienced a rupture in a fuel tank leading to a catastrophic loss of the spacecraft. In other words, it had blown up.

Alan thought to himself, "If we've lost the spacecraft, this entire fourteen-year-long project, and the work of over 2,500 people, will have failed. We won't have learned much of anything about Pluto, and New Horizons will become a poster child for dashed dreams and failure."

THE LINK

As soon as Alan reached the large, mostly windowless office building where the MOC was housed, he parked, pushed dark thoughts out of his mind, and went in to get to work. The MOC looks very much like you'd expect a spacecraft control center to look, just like in *Apollo 13* or other space movies: dominated by the glow of giant projection screens along the walls, and rows of smaller computer screens at consoles.

Throughout the nine long years of travel toward the ninth planet, the radio link to New Horizons was the lifeline that allowed its team to contact and control the craft and to receive spacecraft status and data from its observations. As New Horizons kept going farther to the outer reaches of the solar system, the time delays to communicate with it increased, and the link had lengthened to what was now a nine-hour round trip for radio signals, traveling at the speed of light.

To stay in touch, New Horizons depends, as do all long-distance spacecraft, on a largely unknown and unsung marvel of planetary exploration: NASA's Deep Space Network. This trio of giant radio-dish complexes in Goldstone, California; Madrid, Spain; and Canberra, Australia, seamlessly hands off communication duties between one another as the Earth rotates on its axis every twenty-four hours. The three stations are spread around the world so that no matter where an object is in deep space, at any time at least one of the antenna complexes can point in its direction.

But now . . . the DSN had lost contact with one of their most precious assets, New Horizons.

Alan scanned his badge on the way through building security and arrived in the MOC. Inside, he looked immediately for Alice Bowman, the mission's coolheaded and enormously competent fourteen-year veteran Mission Operations Manager (hence her nickname: "MOM"). Alice led the Mission Control team that maintained communications with and controlled the spacecraft. Alice was huddled with a small group of engineers and mission operations experts in front of a computer screen displaying the ominous message "OUT OF LOCK."

Their calm attitude was reassuring, but it struck Alan that they seemed pretty relaxed, considering the stakes at hand. In fact, as he probed them with questions he learned they were already developing a working hypothesis about what might have happened.

At the time of the signal loss, they knew the spacecraft was programmed to be doing several things at once, a stressful condition on its main computer. Perhaps, they surmised, that computer became overloaded. In simulations, this very same suite of tasks had not been a problem for the identical computer on the spacecraft mission simulator at the MOC. But perhaps something on board the spacecraft was not exactly the same as in the simulations.

If the onboard computer had become overwhelmed with tasks, they surmised, it could have decided to reboot itself. Alternately, it may have sensed a problem and turned itself off, automatically switching authority to its backup main computer aboard New Horizons.

Either one of these alternatives would be good news, meaning that the spacecraft was still alive, and that the problem was fixable. In either scenario New Horizons would have already woken up and radioed home a signal informing them of its status. So if either of the two scenarios was correct, they should hear from their "bird" in about an hour to an hour and a half, once the spacecraft had automatically completed its initial recovery steps. Alice and her team seemed confident that one of these computer problems was the explanation, and after so many years of flying New Horizons, Alan was inclined to give them the benefit of the doubt. Yet, if they did not hear anything—if that hour and a half went by without a signal—it would mean they were without a good hypothesis for what had occurred, and quite possibly without their spacecraft for good.

As more mission staff started to arrive to help address the unfolding emergency, Alan set up shop in the Situation Room, a fishbowl glass conference room that looked out on Bowman's New Horizons mission control room. Glen Fountain also arrived. Soon it became clear that the recovery from this would be involved, and that the team members might be settling in for a long haul—several days of over-nighters to resolve the problem and get the impending flyby back on track.

If this were an orbiter, or a rover safely on an alien surface, the team could take its time to analyze the problem, make recommendations, try different courses of action. But New Horizons was a flyby mission. The spacecraft was rushing toward Pluto at over 750,000 miles per day—more than 31,000 miles per hour. Back to working order or not, it would fly by the planet on July 14, never to return. There was no stopping New Horizons as they sorted the problem out. There was only one shot at getting the goods at Pluto—New Horizons had no backup, no second chance, no way to delay its date with Pluto.

There is a phrase from World War I describing warfare as "months of boredom punctuated by moments of terror." The same applies to long spacecraft missions. And it was a long and frankly terrifying hour as they awaited the hoped-for signal to return from New Horizons.

Then, relief: At 3:11 P.M., 1 hour and 16 minutes after the space-craft signal had been lost, signals returned and a new message appeared on mission control computer screens: "LOCKED."

Alan took a deep breath. The hypothesis that the engineers had formulated must have been correct. The spacecraft was talking to them again. They were back in the game!

Back in the game, yes, but they were still not out of the woods. It was going to take an enormous amount of work to get the spacecraft back on schedule for the flyby. First they had to get it out of "safe mode"—the state the spacecraft goes into when it senses a problem, where every noncritical system is shut off. But there was much more to do to restore the flyby than just that. All the computer files that had been meticulously uploaded since December to support the coming explo-ration of Pluto would have to be reloaded to the spacecraft before the flyby operations could begin. This would be weeks of work under nor-mal circumstances; but they didn't have weeks, they had ten days until New Horizons reached Pluto and only three days until the start of the critical data taking for closest approach, when all of the most valuable scientific observations would be made.

Bowman and her team got to work right away, and the task proved daunting. After they got the spacecraft out of safe mode, they would need to command it to switch from the backup onto the main computer—something they had never had to do before, and then they had to reconstruct and retransmit all of the files that were to orchestrate operations during the flyby. And all of this had to be tested on the mission simulator before anything could be sent to the spacecraft to make sure it would work. And it had to be perfect: if they missed even just one crucial file or used the wrong version of it, much of the flyby they had spent so many years planning could be lost.

The clock was ticking. The first science observations of the close flyby—the most crucial observations at the heart of the mission—would be made just 6.4 days out from Pluto, on Tuesday. That 6.4 days was set by the length of a day on Pluto, one full rotation on its axis, meaning Tuesday was the last time they would ever see large parts of the planet before the flyby. If things were not back on track by then, there would be whole areas of the planet New Horizons would miss mapping—forever.

Could they get the spacecraft back on the timeline by then? Alice and her team created a plan and thought they just might be able to pull it off—assuming no new problems were encountered or were generated by mistakes they might make during the marathon, sleepless recovery effort they were embarking on.

Would it work? Or would they fail? As Alan said that afternoon, if you were on the mission team and weren't religious before this happened, you were probably becoming religious at this point. Time would tell, and so will we, but first let's tell the story of New Horizons, and how it came to reach this point.

I

DREAMS OF A GRAND TOUR

This book tells the story of a small but sophisticated machine that traveled a very, very long way (3 billion miles) to do something historic—to explore Pluto for the first time. It achieved that goal through the persistence, ingenuity, and good luck of a band of high-tech dreamers who, born into Space Age America, grew up with the audacious idea that they could explore unknown worlds at the farthest frontier of our solar system.

The New Horizons mission to Pluto had many roots. They reach back to the astonishingly difficult discovery of Pluto in 1930. They then extend, over half a century later, to the delightful discovery of a host of other worlds orbiting at the edge of our planetary system, and to an underdog proposal to NASA by a determined team of young scientists bent on historic exploration and new knowledge.

Scientists don't necessarily believe in destiny, but they do believe in good timing. So we begin in 1957, the year that the first spacecraft, called Sputnik, was launched into Earth orbit.

KICKING TO GET STARTED

Sol Alan Stern arrived on Earth in New Orleans, Louisiana, in November 1957, the first of three children born to Joel and Leonard Stern. His parents say it was a very easy pregnancy, except for the final few weeks. Then he suddenly began kicking, like crazy. Alan's father maintained, years later at his son's fiftieth birthday party, that Alan had apparently been hearing people talking about the launch of Sputnik, and was clearly impatient to get out and get going to explore space.

Alan grew up interested in science, space exploration, and astronomy, from his earliest days. He read everything he could get his hands on about space and astronomy, but eventually ran out of library books—even in the adult section.

When Alan was twelve, he watched newsman Walter Cronkite on television describing one of the early Apollo landings while holding up a detailed NASA flight plan. "You couldn't actually read it on TV," said Alan, "but you could see it ran hundreds of pages and was filled with all kinds of detail, with every activity scripted, minute by minute. I wanted one, because I wanted to know how space flight was really planned. I thought 'If Walter Cronkite can get one from NASA, then I can get one too.'"

So Alan wrote to NASA, but when told he wouldn't be receiving a copy because he wasn't an "accredited journalist," he decided to double down and fix that issue. Over a year, he researched and wrote by hand a 130-page book. The title was "Unmanned Spacecraft: An Inside View," which—as Alan is the first to note—was "a pretty funny title for a kid who was entirely on the outside and learning as he went."

But it worked. Not only did Alan receive a whole set of Apollo flight plans from NASA, he ended up being taken under the wing of John McLeish, the chief NASA public affairs officer in Houston, often heard narrating Apollo missions on TV. In fact, McLeish began sending Alan a steady stream of Apollo technical documents: not just flight plans, but command-module operation handbooks, lunar-module surface procedures, and much more. Alan became hooked on a space career, but knew he'd have to study for a decade to get the technical skills to join the space workforce after college.

THE GRAND TOUR

Around the same time that John MacLeish was befriending him, Alan also got hold of the August 1970 issue of *National Geographic*, with a cover depicting Saturn as it might appear from one of its moons. The painting, showing the giant, ringed planet cocked at an angle, floating against the black of space over a cratered, icy, alien landscape, seemed at once both realistic and utterly fantastic. The cover story, "Voyage to the Planets," is something that many planetary explorers of Alan's age remember paging through as kids. It contained a level of magic—robotic spaceflight—that today would be found in Harry Potter.

The article described how in the decades to come, NASA planned to launch a series of robotic spacecraft that would explore all the planets and transform knowledge of them from science fiction fantasies into actual photographs of known worlds.

The exploration of the solar system was portrayed as an ongoing sequence of journeys. The article was accompanied by profiles of the first generation of planetary scientists—Carl Sagan among them—who conceived, launched, and interpreted the data from those first voyages. By 1970, NASA had managed to launch only seven spacecraft beyond Earth to reach other planets—three to Venus and four to Mars. These first interplanetary crossings had all been "flybys," missions which simply sent a spacecraft zooming past a planet, with no ability to slow down to orbit or land, gathering as many pictures and other data as possible during a few hours near closest approach. (Note: we say "simply," but, as the following pages of this book illustrate, there is actually nothing simple about it.)

That *National Geographic* article described how the 1970s promised to be "the decade of planetary investigation," with an ambitious list of planned and hoped-for NASA missions that would open up the rest of the solar system to humanity. First, in 1971, would be a pair of orbiters to Mars. Next would be the first missions to the immense uncharted realm of what was then called the outer solar system, as Pioneer 10 and 11 would reach Jupiter in 1973 and 1974 and then travel on to reach Saturn in the distant year of 1979.

Shortly after, Mariner 10 would make the first visit to Mercury, traveling there by way of Venus, where it would make the first ever use of a "gravity assist," a nifty trick that has since become indispensable for getting around the solar system. In a gravity-assist maneuver, a spacecraft is sent on a near-miss trajectory to one planet, which pulls it in and then speeds it toward its next target. It seems too good to be true—like getting something for nothing, but it's not—the equations of orbital mechanics do not lie. For the planet, the tiny loss of orbital speed it trades with the spacecraft has no meaningful effect, but the spacecraft gets a whopping shove in just the right direction. Pioneer 11 was slated to use this same trick during its planned flyby of Jupiter, allowing it to then go on to Saturn.

If all these missions were successful, then before that decade was out, spacecraft from Earth would have visited all five planets known to the ancients—Mercury through Saturn. And what's more, Pioneer 10 and 11, sped up from their close encounters with Jupiter and Saturn, would be racing outward with enough velocity to eventually escape the Sun's gravitational hold entirely, becoming the first human-built artifacts to leave our solar system (along with their uppermost rocket stages).

And then what? There would still be three other planets left to explore, but at the vast orbital distances of Uranus, Neptune, and Pluto it would take an impossibly long time to reach them. Unless . . .

The *National Geographic* article described an ambitious plan to launch a "grand tour" mission that could use multiple gravity assists to visit each of these planets. In theory, a spacecraft could be launched outward toward Jupiter, relayed toward Saturn, and then again relayed successively to each more-distant world. Such a mission would allow all the planets, even distant Pluto, to be reached in less than a decade, rather than the multiple decades such a journey would otherwise take.

But this trick cannot be attempted at any random time, even in any random year or century. The planets, each one on its own orbit around the Sun, need to be arranged in just the right way, like beads strung on an arc, stretching from Earth to Pluto. Like a secret passageway appearing only briefly every couple of centuries, the motions of the planets line up to create such a conduit only once every 175 years.

It just so happened that one such rare opportunity would soon present itself, and it was dubbed the "Grand Tour." Using it, a spacecraft launched by the late 1970s could quickly travel all the way across the solar system, visiting every outer planet in turn and arriving at Pluto by the late 1980s. It was fortuitous that at that moment in history, in the late twentieth century, when humans had just figured out how to launch spacecraft to other worlds, such a rare chance would be coming around.

There were lessons here for a young reader: The laws of physics can be our friends. They can be used to achieve things that would otherwise be beyond reach. And sometimes things line up just right to provide opportunities that, if not seized, won't come around again for a very long time.

That *National Geographic* was illustrated with early spacecraft photographs of Mars and Venus, and artists' depictions of the planets as yet unexplored. It also contained a table summarizing the known facts about all nine known planets, and one planet stood out from the others as completely mysterious. In the column for Pluto, most of the boxes were filled in with just question marks. Only the details of its vast and distant orbit (taking 248 Earth years to complete one of its own) and its length of day (spinning on its axis once every 6.4 Earth days) were given. Number of moons? Unknown. Size? Unknown. Atmosphere? Surface composition? Both also unknown. There was nothing to give us much of a clue about what it might actually be like on Pluto. Alan remembers reading that article and seeing that table, and thinking about spaceships one day exploring mysterious Pluto, the most distant unknown of all the planets.

VOYAGERS

Back then, most interplanetary missions launched as pairs of space-craft, to guard against the possibility that one might fail. There was good logic in that, because the cost of building a second, identical spacecraft is much reduced by borrowing the design and much of the planning for the first. For example, Mariner 9, the Mars orbiter that finally revealed the "Red Planet" in all its detail and glory was suc-cessful. But its twin Mariner 8 ended up crashed beneath in the At-lantic Ocean due to rocket failure. A similar fate had met Mariner 1, though Mariner 2 made it to Venus, and Mariner 3 had failed, but Mariner 4 got to Mars.

NASA's planned grand tour of the giant planets included two pairs of identical spacecraft that would visit three planets each. One pair, to be launched in 1977, would fly by Jupiter and then be ricocheted on to Saturn and Pluto. The other pair would launch in 1979 to visit Jupiter, Uranus, Neptune, and Pluto. The grand tour would complete what Carl Sagan referred to as "the initial reconnaissance of the Solar System."

It was a wonderful plan, but sending four spacecraft to each visit three planets was just too expensive. The projected cost to design, build, and fly this mission, lasting well over a decade and traveling much farther than any spaceflight in history, was more than $6 billion of today's dollars. Sadly, at that time NASA's budgets were falling, and in that environment such an expensive mission was a nonstarter. The grandiose grant tour was canceled before it ever got off the draw-ing board.

Recognizing that the opportunity would not come again in their lifetimes, the science community scrambled to reduce cost and rescue the grand tour, producing a scaled-down version called the "Mari-ner Jupiter-Saturn" mission, with the more modest goals of exploring only the two largest and closest outer-solar-system planets: Jupiter and Saturn. This twin-spacecraft mission, at just under $2.5 billion in today's dollars, was approved in 1972. A contest was held to formally name the spacecraft, and they were christened Voyager 1 and 2 just months before their launches in August and September 1977.

Although the original grand tour had been canceled, the Voyager 1 and 2 launch dates and trajectories were cleverly chosen to enable the craft to keep going after Saturn, using gravity assists to reach all the other planets. The nuclear power source was also designed with enough energy to fly the spacecraft for many years after the "primary mission." So, potentially, these craft could continue on to Uranus, Neptune, and Pluto if funds could later be found to pay for their extended flights.

The Voyager mission would be considered a complete success if it just succeeded in exploring the Jupiter and Saturn systems. Yet its designers planned that—with luck, and future resources they couldn't count on—it just might be possible to keep it going for years longer and billions of miles farther, completing all of the grand tour's objectives after all. And indeed, the Voyagers ultimately did just that. Launched in the late 1970s, each completed its primary mission at Saturn by 1981, and both are still operating today—four decades after launch. Voyager 2 traveled in the direction of Uranus and Neptune, but the wrong direction to reach Pluto, but Voyager 1 headed in the right direction.

So why didn't Voyager 1 go on to Pluto? One of the big prizes, and one of the official metrics for success for Voyager, was the exploration of Saturn's unique and enigmatic, giant moon, Titan. As the only moon in the solar system with a thick atmosphere, even thicker than Earth's, and like the air we breathe made mostly of nitrogen, it naturally stood out as a place scientists wanted to know better. Titan also possessed hints of some interesting organic chemistry (the kind of chemistry, involving carbon, that on Earth enables life to exist), and its atmosphere was known to include the carbon-containing gas methane. This had been discovered in 1944 by astronomer Gerard Kuiper, one of the founders of modern planetary science and someone whose name we'll see again soon.

There was a problem, though, and Titan forced a difficult trade-off. Voyager 1 could only do a really good job of investigating Titan if it made a close flyby immediately after flying by Saturn. Executing such a maneuver would pull the spacecraft permanently off the grand-tour trajectory, flinging Voyager 1 toward the south, veering sharply out of the plane of planetary orbits. This post-Saturn flight direction would make a continuing journey outward to Pluto impossible. At the time, no one could really argue successfully that Voyager 1 should skip Titan. It was a body relatively near at hand compared to Pluto, and scientists knew Titan was fascinating. By contrast, the risky, five-year journey onward to distant Pluto, a body about which so little was known that no one could say it would be worth the effort. Picking Titan over Pluto was a good, and logical, choice. And even today, no one regrets this decision, especially now that Titan has proved to be a world of wonder with methane clouds, rainfall and lakes, and vast fields of organic sand dunes—truly one of the most enticing places ever explored. It was indeed the right decision, but it also closed the door on humanity's chance for a visit to Pluto in the twentieth century. If Pluto were ever to be visited, it would be left for another time, and another generation.

SCHOOL DAYS

Alan finished college, at the University of Texas, in December 1978. Just as Voyager 1 was approaching Jupiter, in January 1979, he then started grad school in aerospace engineering. His fascination with space exploration continued, but he did not see himself becoming a scientist. Even today he remembers hearing about the decision for Voyager 1 to study Titan rather than attempt the longer, riskier journey to Pluto. "I remember thinking back then, 'They made a smart choice, but it's too bad—we'll probably never have the chance to see Pluto.'"

Alan maintained a keen interest in the way spacecraft missions work, but his master's program, with a focus on orbital mechanics, was strategically designed to build a résumé that would enable him to be selected by NASA's astronaut program. What would be the right next move for that?

Alan wanted to show NASA he was versatile, so he went for a second master's in another field, planetary atmospheres. The choice turned out to be pivotal. Alan recalls:

> There was a young, hotshot planetary research professor at Texas who also wanted to be an astronaut, Larry Trafton. He had come out of Caltech and had made some pretty big discoveries. He also had a reputation for rigor and toughness. I remember going to Trafton's office and knocking on the door and feeling very intimidated by his reputation, but telling him I would work for free if he had any ideas for a project we could do together. He told me about a paper he had just written about Pluto that made some calculations about the behavior of Pluto's atmosphere and the high rate that it was escaping into space, which indicated that Pluto should have completely evaporated over the age of the solar system. Of course, this didn't make sense—because Pluto was still there, indicating something else was going on we didn't understand. Trafton just happened to be puzzling over this when I knocked on his door in late 1980, asking for a good research problem to work on. So he said, "Why don't you work on Pluto?" and that eventually became my master's topic. We did some explorations of the basic physics of what Pluto's atmosphere might be like. Very simple computer modeling by today's standards, but illuminating for its time.

Eighteen months later, that second master's in hand, Alan moved to Colorado to work as an engineer on NASA and defense projects for aerospace giant Martin Marietta. But after eighteen months he left for the University of Colorado, where he became project scientist (chief assistant to the project leader, which NASA referred to as the principal investigator) on a space-shuttle-launched satellite to study the composition of Halley's Comet during its once-in-a-lifetime, 1986 appearance. While there he also worked on suborbital science missions and led an experiment to be launched six times on the space shuttle to image Halley's Comet from space—his first instrument as principal investigator.

But as all this was taking place, Alan wondered just how far he could get in this business without a Ph.D. Already married and with a house and a career, he thought he might have missed that boat by not having chosen the doctoral route back at the University of Texas.

Then, in January 1986, tragedy struck. The space shuttle Challenger exploded seventy-three seconds after launch, killing its seven astronauts. The explosion also destroyed both projects Alan had been immersed in the previous three years: the satellite to study the composition of Halley's Comet and his first PI experiment—to image Halley's comet. Beyond the destruction of both his own projects, many of NASA's other plans were also shattered, and the shuttle's future was in doubt.

Nearly everyone involved in space exploration back then remembers where he or she was when the shuttle exploded, and some of us still get teary-eyed thinking of Christa McAuliffe, NASA's first "teacher in space" and the others who lost their lives that cold morning in Florida. Many of us were watching the launch live on TV; Alan was at Cape Canaveral with colleagues, watching the launch.

Following the explosion, Alan was devastated. "You couldn't escape it anywhere on TV or the papers. For weeks, even months, I just kept seeing that explosion over and over again in the media." The experience caused him to rethink where his life and career were headed. NASA's next two planetary missions, Magellan to Venus and Galileo to Jupiter, both orbiters which were supposed to launch on the shuttle, were temporarily shelved. So were virtually all of NASA's other science missions. Concluding that nothing much new would be happening in space exploration until the end of the decade when the shuttle would fly again, Alan decided to go back to graduate school and get that Ph.D.

So Alan entered an astrophysics Ph.D. program at the University of Colorado in January 1987, exactly one year after the Challenger explosion. There he did his dissertation research on the origin of comets. But Pluto had touched his life. It had provided his first real taste of scientific research, and he did start to wonder, even in graduate school in the late 1980s, about the possibility of sending a dedicated mission there. Why hadn't NASA thought more about that?

Alan also realized that, by taking a somewhat circuitous route to a Ph.D., he had fallen a few years behind his peers who had pushed straight through. Others his same age had graduated or been in grad school in time to be involved in the excitement of the Voyager project. Had he missed out on the last opportunity to explore new planets for the first time? Not if he could be involved in a mission to Pluto.

When he first raised the matter with senior planetary scientists, the response was not encouraging. Alan:

> I think I'm different from most people in our field in the extent to which I'm really inspired by exploration itself, independent of the science. When I was working on my Ph.D., I first started floating the idea of a Pluto mission, saying "We learned so much at Neptune. Why don't we do a Pluto mission?" It was disappointing to me to learn that senior scientists insisted that a mission to Pluto wasn't justified simply for its exploration value.

Right there, Alan encountered a basic disconnect between the way NASA actually makes exploration decisions, and the way its efforts are often portrayed to the public. When NASA does public outreach, it often stresses the excitement and intrinsic value of exploration. It's all about "We are boldly going where no one has gone before."

But the committees that assess and rank robotic-mission priorities within NASA's limited available funding are not chartered with seeking the coolest missions to uncharted places. Rather, they want to know exactly what science is going to be done, what specific high-priority scientific questions are going to be answered, and the gritty details of how each possible mission can advance the field. So, even if the scientific community knows they really do want to go somewhere for the sheer joy and wonder of exploration, the challenge is to define a scientific rationale so compelling that it passes scientific muster.

Alan remembers how, in the late 1980s, "Somebody much more senior told me, 'You will never sell going to Pluto to NASA as exploration. You have to find a way to bring the scientific community to declare it is an important priority for the specific science that such a mission will yield.'"

DISCOVERING PLUTO—1930

Of all the classically known planets, Pluto was not just the farthest and the last to be explored, it was also the most recently discovered— within the lifetime of many people who are still alive. Its discovery in 1930 by Clyde Tombaugh, a Kansas farm boy with no formal technical training, is a classic tale of stubborn perseverance leading to a big payoff.

Clyde, born into a hardscrabble Illinois farming life in 1906, grew up fascinated by thoughts of other worlds. "One day, while in sixth grade," he wrote in an autobiographical sketch, published in 1980,* "the thought occurred to me, What would the geography on the other planets be like?" As he grew up, his family moved to a farm in Kansas, where he studied the sky diligently with a 2¼-inch telescope his dad purchased for him from the Sears, Roebuck catalog. He studied astronomy on his own, grinding lenses to build new telescopes and making careful drawings of the markings he observed on Jupiter and Mars. He read everything he could find in the local library relating to astronomy and planets, and he followed the debates about the controversial "canals" on Mars "discovered" and promoted by the wealthy and charismatic Boston astronomer Percival Lowell. He also read about Lowell's prediction of an undiscovered planet beyond the orbit of Neptune.

Lowell had carefully examined Neptune's orbit and concluded that

---

*Clyde Tombaugh and Patrick Moore, *Out of the Darkness* (Harrisburg, PA: Stackpole Books, 1980).

some irregularities in its motion betrayed the slight gravitational pull of a distant ninth planet. Clyde read about the observatory Lowell founded on a mountain above Flagstaff, Arizona. He fantasized that he, too, might someday go to college and become an astronomer, but his life felt many worlds apart from all that. Times were not good, and he couldn't imagine his family ever having the money for him to leave the farm and pursue these dreams.

Still, ever hopeful, he mailed some of his best sketches of Mars to the astronomers at Lowell Observatory. One day in late 1928, to his amazement, he got a letter back from the observatory director, Dr. Vesto Slipher. They were hiring an assistant and wanted to know if he was interested in the job.

You bet he was! In January 1929, with little more than a trunk full of clothes and astronomy books, and some sandwiches his mother had made him for the journey, Clyde boarded a train west, to Arizona. Three weeks shy of his 23rd birthday, excited but a little sad to be leaving the family farm, he watched the scenery change from flat Kansas farmland to dry desert to pine-filled forests as the train chugged out of Kansas and up into the Arizona mountains. That train also carried him—though he didn't know it—into history.

When Clyde arrived he learned he'd been hired to use the brand-new thirteen-inch telescope to renew the search for "Planet X." In this amazing assignment he would be taking up a quest that had been started by the famous Percival Lowell. Lowell had died in 1916 without ever finding this prey; now it was Clyde's job to resume the search.

The new telescope built for hunting down Planet X was better than what Lowell himself had used, and the observatory's location, at seven thousand feet in the northern Arizona mountains, provided dark, dry skies. The search work assigned to Tombaugh was painstaking. He spent night after night in the unheated telescope dome throughout the icy winter, taking photographic plates, one after the other, each of tiny portions of the sky in the region where orbital calculations predicted the new planet might be.

The planet Clyde was looking for was expected to be so faint (thousands or even tens of thousands of times fainter than the eye could see), that each photographic plate had to be exposed for more than an hour, while he guided the telescope carefully to compensate for Earth's rotation and to keep the stars stationary in the frame. Each frame was studded with thousands of stars, occasional galaxies, and many asteroids, and even occasionally comets.

How would Clyde even know that any given point of light was a planet? The key was photographing the same little spot for several nights in a row to detect his faint prey moving against the stars at just the right rate to indicate it orbited beyond Neptune. To analyze the images, he used a device, state of the art at the time, called a "blink comparator" that let him flash between images from successive nights. The background stars would remain still as he blinked between frames, but a planet would show movement.

It's hard to really overstate how tedious and demanding this must have been. Today every part of this work would be done by computer, but it was all done manually back then. So Clyde went to the telescope every night when the weather permitted and the full moon was not washing out the deep darkness. He lived by the twenty-eight-day lunar cycle, using the downtime afforded by the full moon, when the sky was too bright to photograph the sky for his faint, hoped-for planet, to develop the prints and laboriously examine them, blinking between frames to check first one spot and then the next.

Success was far from guaranteed. Some senior colleagues told him that he was wasting his time; that if there were any more planets, they would have already been found in previous searches. No wonder Clyde suffered bouts of low morale and self-doubt. But still he kept going.

After nearly a year of arduous hunting, on January 21, 1930, the sky was clear and, in his systematic sweep across the sky, his search took him into an area within the constellation of Gemini, the "twins." It turned out to be a horrible night because of an intense wind that came up, shaking the telescope and nearly blowing the door off the hinges. The images he took were so blurry that they seemed useless, but it turned out that—though he didn't know it at the time—Clyde had actually photographed his long-sought target, Lowell's Planet X.

Because the weather conditions on the twenty-first had been so poor, Clyde decided to photograph the same region again on January 23 and 29. It was a good thing he did.

A few weeks later, on February 18, when the nearly full moon once again made searches for faint targets impossible, he set to work blinking between his January images, looking for something that moved at just the right rate to indicate it was at a greater distance than any of the known planets. He had found, by trial and error, that alternating back and forth between the frames at a rate of about three times per second worked best. On one of his January plates, he saw something that matched what he was looking for. A faint, tiny speck was dancing back and forth by about an eighth of an inch—just the right amount to be out beyond Neptune. "That's it!" he thought to himself. Clyde:

> A terrific thrill came over me. I switched the shutter back and forth, studying the images. . . . For the next forty-five minutes or so, I was in the most excited state of mind in my life. I had to check further to be absolutely sure. I measured the shift with a metric rule to be 3.5 millimeters. Then I replaced one of the plates with the 21 January plate. Almost instantly I found the image 1.2 millimeters east of the 23 January position, perfectly consistent with the shift on the six-day interval on the discovery pair. . . . Now I felt 100 percent sure.*

At that moment, Tombaugh knew he had bagged his quarry. He also knew that he was the first person to discover a new planet in decades.† That minuscule pale dot, hopping back and forth like a flea on a dark plate surrounded by a forest of stationary stars, was the first glimpse of a place never before spotted by human eyes.

---

*Clyde Tombaugh and Patrick Moore, *Out of the Darkness* (Harrisburg, PA: Stackpole Books, 1980).
†Tombaugh came after William Herschel, who discovered Uranus in 1781, and Johann Galle, who found Neptune in 1846. Galle often shares credit with Urbain Le Verrier, who predicted Neptune's position exactly based on calculations made from irregularities in the orbit of Uranus, and so told Galle where to look.

There was another planet out there! And for a few long minutes, Clyde Tombaugh was the only person on Earth who knew. Then, sure of his find, he walked slowly down the hall to tell his boss. As he stepped down the corridor, he weighed his words. In the end, he walked into the observatory director's office and said, simply, "Dr. Slipher, I have found your Planet X."

Slipher knew how careful and meticulous Clyde was. Clyde had never made such a claim before, and this was not likely to be a false alarm. After Slipher and another assistant inspected the images and concurred with Tombaugh's assessment, they agreed with him but resolved to keep it tightly guarded, telling only a few colleagues at the observatory. Meanwhile, Clyde made follow-up observations to further confirm the discovery and to learn more details about what kind of object it was and how it was moving. A false claim would be devastating.

They spent more than a month checking and rechecking on the new planet and its path in the sky, confirming the calculations that showed it was farther out than Neptune. The planet passed every test they had, appearing in every new image and moving at just the right speed. They also spent the month searching for moons around the new planet (they found none) and trying—with a more powerful telescope—to see it as an actual disk, rather than just a point, so they could estimate its size. They could not, which suggested their planet was small.

Finally, sure of their find, on March 13, 1930, which was both the 149th anniversary of the discovery of Uranus and what would have been Percival Lowell's 75th birthday, they announced their discovery.

In no time at all, the sensational news spread around the world. The New York Times ran the banner headline: NINTH PLANET FOUND AT EDGE OF SOLAR SYSTEM: FIRST FOUND IN 84 YEARS, and the story was run by countless other papers and radio broadcasts.

The discovery was a huge feather in the cap of Lowell Observatory, which soon felt pressure to choose a name for the new planet quickly, before someone else did. Percival Lowell's widow, Constance, who had previously engaged in a ten-year battle to rob the observatory of the endowment her late husband had provided for the planet search, now insisted that the planet be named "Percival," or "Lowell." Then she wanted it to be called "Constance," after herself. Naturally, nobody wanted that, but it was a tricky situation for the observatory, which was still financially dependent on the Lowell family.

Meanwhile, more than a thousand letters arrived suggesting names for the new planet. Some were serious suggestions, based on mythology and consistent with the names of the other planets. Among them were Minerva, Osiris, and Juno. Others suggested modern names like "Electricity." Still other submissions were bizarre or unlikely: a woman from Alaska sent in a poem to support her contention that the planet should be called "Tom Boy," in honor of Tombaugh. Someone from Illinois volunteered that the planet ought to be named "Lowellofa," after Lowell Observatory in Flagstaff, Arizona. And a man from New York suggested "Zyxmal," because it was the last word in the dictionary and thus perfect for "the last word in planets."

But it was Venetia Burney, an eleven-year-old school girl in England, who suggested the name "Pluto," after the Roman ruler of the underworld. Her grandfather mentioned Venetia's idea to an astronomer friend, who in turn sent a fateful telegram to Lowell Observatory, saying:

> Naming new planet please consider Pluto, suggested by small girl, Venetia Burney, for dark and gloomy planet.

Clyde and more-senior astronomers at Lowell liked the name and proposed it to the American Astronomical Society and the Royal Astronomical Society of England, both of whom also liked it. The Lowell astronomers thought that the name "Pluto" was perfect, not only because it fit with the convention of naming a planet after an appropriate classical deity, but also because the first two letters in Pluto were *PL*, which could also serve to honor their founder and benefactor: Percival Lowell.

2

THE PLUTO UNDERGROUND

THE PLUTOPHILES

With the advent of actual planetary exploration in the 1960s, the planets, once merely points of light glimpsed vaguely through telescopes, became real worlds to be reconnoitered and studied with powerful new tools and techniques, including many borrowed from the study of our own home planet, the Earth. Planets have rocks and ice, landforms, weather, clouds, and climate. So the effort to figure out the planets drew in geologists, meteorologists, magnetospheric experts, chemists, and even biologists. Given its complexities, it especially attracted adventurous scientific types who were up for novel interdisciplinary challenges. In the process, a new and distinct field was born: planetary science.

Of all the planets, Pluto—the farthest and the hardest to reach— remained the most obscure and cryptic, and the most difficult to study. But planetary scientists like observing challenges, and they like puzzles, and Pluto provided plenty of each. There's a drive in scientists to know, and there's a drive to contribute to the knowing. And with so many mysteries to attack at Pluto, a determined sub-community of scientific Plutophiles developed. Hungry for new information, they employed the most sophisticated telescopes and other advanced tools to puzzle out what they could from so far away, back here on Earth.

One of the first things discovered about Pluto, with the primitive tools available immediately after its 1930 discovery, was the size and shape of its orbital path, which, compared to any other then-known planet, was both huge and genuinely strange. One thinks of the Sun as unimaginably far away from Earth. And it literally is—93 million miles away. When we hear that, our brains have trouble comprehending this as anything other than a really, really big number. So it is common to use analogies to grasp it, for example reducing the Earth to the size of a basketball puts the Sun at 5.5 miles away! But Pluto orbits at an average distance some forty times farther from the Sun than Earth, which means that on this same basketball-Earth scale, Pluto is located 220 miles away!

At that great distance, the Sun's gravitational hold is much weaker and planets orbit much slower. As a result, Pluto takes 248 years to make it just once around the Sun. Think about this: Roughly one Pluto year ago, Captain Cook was just setting sail on his first voyage from England, and only a little over one-third of a Pluto year ago Clyde Tombaugh was first eyeing the dot, Pluto, jumping in the frames of his blink comparator at Lowell Observatory.

Pluto's path is also highly elliptical (i.e., noncircular), more so than any closer planet. Because of this, from the time of its discovery to the late 1980s, Pluto had been slowly heading inward, ever closer to the Sun and to Earth. In the 1950s its increasing brightness, along with the development of new tools for precisely measuring the brightness of astronomical objects, allowed the first detailed measurements of Pluto's "light curve"—that is, the way its brightness changes as it rotates on its axis. What jumped out from this analysis was a regular pulsing, a brightening and dimming, that occurs precisely every 6.39 Earth days. By spotting this slow, steady rhythm, scientists had discovered the length of Pluto's day. Whereas Earth takes twenty-four hours to rotate once, Pluto spins at a comparatively stately pace, taking about

6.4 times as long between each successive sunrise—more slowly than any planets except Venus and Mercury.

As technology improved in the early 1970s, planetary astronomers also succeeded in recording the first crude spectrum of Pluto, meaning they succeeded at determining its brightness as a function of wavelength. This revealed that its overall color is a reddish hue.

Then in 1976, planetary astronomers at a mountaintop observatory in Hawaii discovered the subtle spectral fingerprints of methane frost (frozen natural gas!) on Pluto's surface.* This provided the first evidence that Pluto's surface is made out of truly exotic stuff. The team that discovered methane on Pluto realized that the discovery also had an important implication for the size of Pluto, which was then unmeasured. All scientists knew at that time was how much light Pluto reflected in total. From this they could derive its size if they knew—or assumed—how reflective the surface was. Because methane frost is bright and reflective, its discovery meant that Pluto was likely small.

Next, in June 1978, James "Jim" Christy, an astronomer at the U.S. Naval Observatory, observed "bumps" on some of his images of Pluto. Were these real features? Or were they imperfections in the images? Christy noticed that the stars in the same images had no such bumps, only Pluto did. So he analyzed the time between appearance of bumps and found a familiar period—6.39 Earth days, the same as Pluto's rotation period! Other observers, alerted by Christy, found something similar. He had discovered that Pluto has a moon that circles it closely in an orbit that has precisely the same period as Pluto's day. He later named it Charon (most often pronounced "Sharon"), after the mythical Greek ferryman who carried the dead to the Plutonian underworld. Choosing this name also cleverly allowed

---

*The classic paper reporting on this discovery and making other clever inferences about size and reflectivity was "Pluto: Evidence for Methane Frost," written by Dale Cruikshank, Carl Pilcher, and David Morrison, published in *Science* on November 19, 1976.

Christy to name Pluto's moon to sound like his wife's name, Charlene. After all, scientists are people, too.*

Christy's discovery of Charon turned out to be a mother lode for learning more about Pluto. Careful observations of the changing position and brightness of Charon revealed the size of its orbit. This, and the laws of physics, allowed the elusive question of Pluto's mass to finally be nailed down, and the result was a bit shocking. Whereas Lowell, Tombaugh, and many subsequent Pluto researchers had expected a planet of roughly Earth mass or even higher, Pluto turned out to be only about 1/400th the mass of Earth. Rather than being another giant like Neptune, it was a tinier planet than any previously discovered.

And the surprises didn't stop there. Compared to Pluto Charon was huge, with a mass almost 10 percent of Pluto's: the pair literally formed a double planet (sometimes also called a binary)—a first in our solar system! A double planet was something completely unknown in planetary science before the discovery of Charon. Pluto, it seemed, became more exotic every time we learned something new about it.

But beyond what Charon immediately taught us, one other aspect of Charon's discovery turned out to be bizarre, yet scientifically convenient. It turned out that when Charon was discovered, its orbit was about to enter into a very unusual geometric arrangement: essentially it was about to point right at us. Yes, that sounds strange—and it is strange. Because the tilt of Charon's orbit remains steady while Pluto makes its slow circuit around the Sun, occasionally, but only briefly, Charon lines up to repeatedly pass (in our view) directly in front of and then behind Pluto. Such a perfect alignment happens for just a few Earth years per 248-year-long Pluto orbit. Amazingly, this chance alignment was just about to begin only a few years after Charon's discovery. So, by pure luck, as seen from Earth, the newly discovered moon and its planet were about to start repeatedly eclipsing each

---

* We should point out that the first analysis of the discovery observations was done by Robert Harrington and James Christy. It was Christy who first noticed the "bumps" and inferred the existence of the moon, but in the scientific literature Christy and Harrington rightfully share credit for the discovery.

other—a scientific bonanza that would teach us many things about the distant planet Pluto and its big moon Charon.

Estimates indicated that these eclipses would start happening in 1985, give or take a few years, and should continue for about six years. During this "mutual event season" such an eclipse should occur every 3.2 days—that is, every half Charon orbit period of 6.4 days. Those eclipses would allow the sizes and shapes of both bodies to be determined for the first time, and would provide many additional clues about their surface brightnesses, compositions, colors, and possible atmospheres. In the pantheon of strange and lucky coincidences, the fact that Charon was discovered just before it was about to swing into place for this mutual event season—after all, another such set of eclipses wouldn't occur for more than a century—is right up there with the grand-tour alignment of the planets appearing just when humans were ready to take advantage of it by mastering spaceflight.

One of the Plutophiles primed for observing these mutual eclipses was a young scientist named Marc Buie, another child of the Space Age who had been deeply and permanently touched by black-and-white television images of people in rockets blasting off in his youth. As a college student, he'd caught the Pluto research bug and never got cured.

Marc then went off to grad school at the University of Arizona in Tucson. Finishing his Ph.D. in 1985, his timing was perfect, because Pluto and Charon were about to start eclipsing each other. Marc:

> We were lucky enough to have discovered Charon right before the mutual eclipses, so we could do this wonderful six-year observing effort, which put Pluto on the docket for every major planetary science conference from then on, and raised Pluto up another notch in the collective scientific consciousness.

In the mid-1980s several observing groups had been watching for the predicted eclipses to start, but nobody knew exactly when the first ones would occur. The first detection of them was observed by a young scientist named Richard "Rick" Binzel in February 1985. Then, once it was known the events had begun and Pluto and Charon started casting shadows on each other every 3.2 days, more observers got in on the game, and they provided a rush of new results, including hints of interesting surface features and surprising differences between the pair of bodies. As Marc put it, due to these eclipses, "Pluto hit the main stage."

THE UNDERGROUND

As the 1980s drew to a close, Voyager 2 was nearing the end of its landmark exploration of the giant planets, culminating in its flyby of Neptune and Neptune's planet-size satellite Triton.

As the end of Voyager's exploration of the four giant planets approached, a group of young scientists, all at the beginning of their careers, and conditioned by the boldness and near-mythical success of Voyager, started dreaming and scheming about how to keep the spirit and the quest of first-time exploration alive, how to go farther still— to explore Pluto.

In 1988, and still a grad student, Alan started to ponder the possibility of sending a spacecraft mission to Pluto. He could see that the first, and in some ways biggest, hurdle to getting such a mission started would be more social and political than technical or scientific.

To succeed, the concept would need a critical mass of support within NASA and the scientific community. Alan knew that the mutual events between Pluto and Charon had yielded many exciting results that painted a new picture of Pluto, showing it to be a highly exotic planet. And the late-1980s discovery of Pluto's atmosphere had added more fuel to that fire. Alan knew that this increasing scientific interest in Pluto might be channeled to support a spacecraft mission there.

But did NASA know this? Not really, it seemed. Pluto never appeared on the short lists of high-priority missions on the reports that came out from influential committees guiding NASA policy.

Yet Voyager's decision not to go to Pluto had come before the discovery of the atmosphere, the discovery of its giant moon, and the growing evidence that it had a complex and varied surface. Pluto had, in the years since the Voyager decision not to visit it, become a lot more enticing.

Moreover, the six-year season of Pluto-Charon eclipses had already revealed a lot, including dramatic surface variations between bright and dark areas on Pluto and the fact that the surfaces of Pluto and Charon are, surprisingly, made out of very different ices. Pluto's surface features frozen methane, whereas Charon's was found to be made of water ice. Continued observations of "occultations"—the temporary dimming of stars when Pluto passed in front of them— had also revealed some weird atmospheric structure, possibly indicating a low-altitude haze, hinting even more at a surprisingly complex planet.

Given all these factors, Alan saw Pluto as the obvious next place to explore, and he sensed that it was a good time to see if he could drum up support for that exploration among planetary scientists.

He wasn't quite sure how to do it, but Alan thought that a good first step would be to gather Pluto scientists together in a highly visible scientific forum. At the time, there were only about one thousand active planetary scientists, and most attended the annual spring and winter meetings of the American Geophysical Union (AGU), where scientists congregate for a week to attend "sessions" of talks organized around different topics. Alan and a few colleagues decided that they would organize a technical session about new discoveries and insights into Pluto, and would propose this to the committee arranging the upcoming spring AGU meeting, in Baltimore in May of 1989.

For this task of organizing and recruiting other scientists, Alan enlisted the help of Fran Bagenal, a young, hotshot British planetary scientist who had made a big impression with her work on the Voyager science team. Fran had just been hired for her first "real job," a junior professorship of the Department of Atmospheric and Planetary Sciences at the University of Colorado, where Alan was finishing grad school and where David would soon become a professor.

Fran was not a Plutophile. In fact, at first she needed convincing that a Pluto session at AGU even made sense. Back then, Fran thought of Pluto as merely a distant curiosity, and it was not the kind of place for her particular genre of scientific wizardry, which involves magnetic fields. To someone who looks at a planet and first sees those magnificent, magnetic structures, what good was a little ice ball like Pluto, a world too small to likely have any magnetic field? Fran:

> To be honest, I didn't think much of Pluto. It was a small object in the outer solar system. It probably didn't have a magnetic field. What kind of interaction with the solar wind was it going to have, and why bother to study it? Was it worth going all that way to visit some battered chunk of ice?

But Fran was a rising star of planetary science, and Alan wanted her on his team. Looking back on it now, she sees her initial involvement as stemming from a combination of Alan's advocacy and the impending need, after Voyager, for inspiring and daring new exploration:

> Alan was recruiting and galvanizing a group of people to work on Pluto. I actually remember thinking, "Oh, Voyager's going to be over; now what do we do?"

So at Alan's urging, Fran attended some Pluto science meetings, and she soon started to catch the Pluto bug:

> There was a transition to my thinking, and that happened before the AGU conference in May of 1989. I found that indeed there were smart people who had been looking at Pluto's atmosphere and surface and finding curious things. So I brought in Ralph McNutt, and we sat down and looked at it and realized there could be quite an interesting interaction with the solar wind. That's how we started to get curious about the physics that could be occurring on Pluto.

Fran and Ralph McNutt had met when they were both grad students at MIT, both working with the Voyager plasma instrument team, studying magnetic fields. After a period working for "the dark side" at Sandia National Laboratories in New Mexico (where nuclear weapons are developed), Ralph ended up back at MIT as a faculty member, where he was during the Voyager flybys of Uranus and Neptune. Ralph recalls:

> At Alan's urging Fran and I submitted an abstract to that 1989 AGU meeting discussing Pluto's possible solar-wind interactions. She gave the talk, and we wrote up a paper about it. We both started to understand that the interaction between the solar wind and Pluto's atmosphere of evaporating methane was actually a really important thing to study. Then I realized that we needed a mission to figure it out. I got on board.
>
> After that, anytime anybody said "Pluto" to me, I would say, "We need to send a spacecraft there. We have to finish the exploration of the solar system, and we need to do it right." I got caught up in the enthusiasm of Fran and Alan and the others who realized Pluto's scientific potential. None of us were going to take no for an answer, and we were all so young that we didn't know any better than to try.

With Ralph's encouragement and collaboration, Fran got involved with Alan in organizing that first ever AGU Pluto session. Then they put the word out to the community of scientists doing research on Pluto to "vote with their feet" by submitting research talks and attending the session to show interest in a possible mission to explore Pluto.

Among those who heard the call was a bright and iconoclastic geophysicist named William "Bill" McKinnon. Bill had just recently been hired as an assistant professor at Washington University in St. Louis. His specialty is planetary geophysics, which means he applies knowledge and techniques used to study the internal structure and movements of the Earth to make sense of the same on other planets and moons. It took some imagination and courage to apply geophysics to a body like Pluto that was still barely more than a point of light. But Bill was fascinated with the origin and geology of icy worlds like the satellites of the giant planets and distant Pluto.

Very tall and a little gaunt, with angular features and long dark hair, Bill looked a little like Frank Zappa, or at least more like someone you'd run into at a Zappa concert than at a planetary science conference. Bill was and is a serious rock-music fan, and he has a dry, slightly dark sense of humor. He's also one of the smartest people anyone could ever meet and one of those lovable nerds with whom you just want to talk science forever.

Bill had authored a pivotal research paper published in *Nature* in 1984—"On the Origin of Triton and Pluto." Decades later, it's still regarded as a classic. The paper refuted a then-common idea for Pluto's origin, that it could have started out as a satellite of Neptune—a twin to Triton—that had escaped into orbit about the Sun. Bill's work modeled all the gravitational and tidal jostling that could have occurred between these bodies, and his convincing and clever calculations showed that the only plausible origin scenario for Triton and Pluto was exactly the opposite. Bill showed the story was one of capture, not escape. Pluto was not a runaway moon of Neptune. Rather, Triton had started off like Pluto, a small, freely orbiting planet of the solar system, but it had been caught by Neptune's gravity and drawn into orbit about that giant planet. His paper concluded, "The simplest hypothesis is that Triton and Pluto are independent representatives of large outer solar system planetesimals."

Throughout the 1980s, McKinnon had continued to work on the origin of Pluto, becoming convinced that if we could ever get there, we'd discover much more than just the random idiosyncrasies of one small world. Pluto, McKinnon realized, could provide a glimpse behind the veil of a whole kingdom of new worlds containing missing information on the construction of our entire solar system. So he certainly did not need to be persuaded about the need for a mission there.

When he learned of Alan's AGU session, he contributed a talk entitled "On the Origin of the Pluto-Charon Binary."

Planning for the AGU session was coming together well—with many research-talk contributions and almost all of the most prominent scientists studying Pluto planning to attend. With this success almost in hand, Alan sensed that it might also be a good time to try to plant the seed for a Pluto mission directly in NASA Headquarters.

So about a month before the AGU members gathered in Baltimore, he requested and was granted a one-on-one meeting with Dr. Geoff Briggs, then NASA's director of the Solar System Exploration Division. It was not your usual meeting request by a graduate student, but having been older than average and having had a previous career as a spacecraft engineer, Stern knew Briggs and leveraged that relationship.

The week before the AGU meeting, Alan visited Briggs in his office at NASA Headquarters in Washington, DC. He told Briggs about the upcoming Pluto session at AGU, about how much great new science there was and about the growing wave of interest in Pluto. He asked Briggs, "With Voyager winding down, why don't we complete the job of exploring the solar system? Would you fund a study of how to do a mission to Pluto?"

Alan was surprised by Briggs's immediate, unhesitating, and positive response: "You know, no one's ever asked me about that before. It's a wonderful idea: we should do that."

DINNER AND A MOVEMENT

Everything starts out somewhere, and big things sometimes start out very small. The origin of what ultimately, much later, would become the almost-billion-dollar New Horizons mission to Pluto, happened one night in May of 1989 in an unremarkable Italian restaurant in Baltimore's Little Italy neighborhood. A large contingent of planetary scientists were in town for the first ever AGU Pluto session, featuring a dozen talks by an impressive coterie of scientists. The session was attended by a hundred or more other scientists and generated good buzz. Alan and Fran, knowing that all the key Pluto people would be there, wanted to strike while the iron was hot. So they arranged for a core group to have dinner together the evening after the Pluto session to discuss how to go forward toward a mission. Alan, Fran, Marc, Ralph, Bill, and nine other scientists were there. None of them could imagine then that something historic was being hatched that evening.

Over meatballs, pasta, and Cabernet, the dinner group started talking about what it might take to send a mission to Pluto. The task was clearly daunting. No Pluto mission was on the drawing board at all, and there were numerous other missions already in line, each waiting their turn to fly. Each had its own constituency of advocates. There were the Mars people, the Venus people, those working on Cassini— the promised (but expensive) big ticket Saturn-orbiter mission—and those who dreamed of returning a sample from a comet. Each had impressive plans for missions that were long overdue. But NASA's budget was so limited that only two new planetary missions had been started in all of the 1980s.

The dinner group knew they were out of their depth, but they shared a conviction that a Pluto mission was an important idea that somehow needed to happen. Alan had been working on a plan of attack, and the group was all ears. He recounted his meeting at NASA Headquarters with Geoff Briggs and the amazingly easy path he found to get agreement for a mission study.

The next question was: How could they rally the planetary scientific community to show NASA that a Pluto mission had broad support? They brainstormed ideas and formed a plan to build cred and buy-in. They scribbled action items on napkins. One was to publish a special issue of the *Journal of Geophysical Research* showcasing the research results from that day's AGU Pluto session. Another was to work to excite the people they knew at NASA Headquarters, to follow up on the idea of sponsoring a mission study. They would also start to recommend Pluto mission supporters for the various committees that advise NASA about planetary mission priorities. And they would organize a letter-writing campaign to cajole colleagues to contact NASA and express support for a mission.

It helped that the Plutophiles were mostly young and at the start of their careers. For these ambitious space nerds in their twenties and thirties, who had grown up on Apollo, Mariner, Viking, and Voyager, the thought of upending the established order with an insurgent Pluto mission was more than a little crazy, and also thrilling. Even if they were tilting at windmills, it would be fun. They loved the idea of defying the odds and bucking the establishment.

It wasn't that night that they started calling themselves the Pluto Underground. In fact, nobody remembers exactly when the name arose, but it began to be used and was fitting in many ways. It started, no doubt, as a play on the term "Mars Underground," a group of passionate scientists and space enthusiasts who had been shaking up NASA with imaginative and aggressive plans for human bases on Mars. The Mars Underground has helped goad NASA into planning a new generation of Mars missions. If anything, though, the "underground" tag was even more apt for the Pluto fanatics. Because a mission to Pluto was, at first, a subversive and unlikely idea, cooked up by a rebel alliance that seemed ill-equipped to take on an empire.

Marc Buie remembers coming away from that Italian dinner with a changed view and an assignment to act:

> I consider that dinner to be a pivotal moment. That was the turning point where it all went from just gee-whiz hallway conversations to a larger, more systemic plan of attack to try to accomplish something. I left with a task: to start a grassroots letter-writing campaign to NASA by other scientists. I went back to my office and I wrote a letter that I sent out to everybody I could think of in the field, young and old, encouraging them to write letters to NASA Headquarters saying, "We really ought to think about going to Pluto."

The letter-writing movement that Marc was assigned was a bit of an end run, going outside the normal NASA advisory process, or at

least gaming it a little. And it was typical of the ad hoc, guerrilla nature of their movement. It also got Marc in hot water. He was a junior
scientist at the Space Telescope Science Institute in Baltimore, where
he was the point person for all the planetary observations on the
Hubble Space Telescope. Marc's "Dear Colleague" letter came to the
attention of his boss, Riccardo Giacconi, the powerful and intimidating (and soon to be Nobel Prize–winning) director of the entire
Space Telescope Science Institute: "He called me into his office and
really lit into me over that campaign and accused me of lobbying. I
said, 'I know what lobbying is and that's not what this is. I should be
able to talk to my colleagues about my scientific interests.'"

Buie's letter-writing campaign generated dozens of appeals to NASA,
so the Pluto Underground's rabble-rousing quickly started to have an
effect. The biggest planetary science meeting every year is the "DPS
meeting," which is shorthand for the Division of Planetary Sciences of
the American Astronomical Society. At the DPS meeting that fall,
NASA held an evening session in which their officials presented
future mission plans to the planetary science community and sought
input. Geoff Briggs stood up and told nearly one thousand planetary
scientists how NASA Headquarters had been flooded with letters
urging it to study a mission to Pluto. As he described it, such strong
community interest took NASA by surprise, and NASA was starting
to take it more seriously.

As a result, right around the time of that DPS meeting, just four
months after Alan had approached him at NASA Headquarters to do
a study, Briggs funded the first official NASA study of a possible
Pluto mission. He asked Alan, fresh out of grad school, and Fran, just
teaching her first class as a professor at Boulder, to be the lead scientists of that study, pairing them with a highly experienced and brilliant NASA engineer a generation older: Dr. Robert Farquhar, who
would manage the study. It was a good sign: Farquhar had a legendary reputation as an innovative and visionary mission designer.

The timing dovetailed fortuitously with another wave then washing
through NASA's planetary mission culture, recognizing that the program needed more small missions.

Given budgetary reality, there just wasn't the money to afford new big missions like Voyager. Why? Galileo and Cassini, NASA's next two big approved missions, each multibillion-dollar giant-planet orbiters, were both experiencing major problems at the time. Galileo had just been launched on its way to orbit Jupiter, but it had issues. The main antenna was broken and the data rate back to Earth had to be drastically reduced, lowering goals and expectations for what it could achieve. Cassini—a Saturn orbiter NASA was then building—was next, but its costs kept ballooning and there was a serious threat of cancellation. These missions were—in the lingo of the profession—"Christmas trees," meaning they were loaded with so many capabilities and scientific instruments, which resulted in those multibillion-dollar price tags.

In the fiscal climate of the early 1990s, any more such missions were simply nonstarters. Instead, Briggs and others at NASA encouraged the development of smaller, cheaper, better-targeted missions with more modest payloads containing far fewer instruments and capabilities, encouraging creative thinking of ways to do ambitious missions with smaller price tags.

Farquhar's Pluto mission study ran for a year, wrapping up in late 1990. It was called "Pluto 350," and it was focused on a small, 350-kilogram spacecraft weighing just about half what Voyager had. The resulting design included a much smaller instrument payload than Voyager's, but with more-compact and modern instruments designed to maximize science per pound, including a camera and infrared spectrometers to photograph and map Pluto's surface, an ultraviolet spectrometer to examine the atmosphere, and a plasma instrument to measure interaction with the solar wind.

Farquhar (who, sadly, died in late 2015, shortly after witnessing the Pluto flyby just months before) was a genius of orbital mechanics and had a legendary knack for finding crafty solutions to get from planet to planet with less fuel than others thought possible, primarily by using clever gravitational assists. One of his innovations that greatly lowered the anticipated costs of the Pluto 350 mission was to plan a launch on a relatively small rocket, a Delta II. A liability of this plan was that it wouldn't have enough velocity to fly straight to Jupiter for the gravity assist toward Pluto. Instead, in Farquhar's scheme, Pluto 350 would first go inward toward the Sun, use gravity boosts from Venus and then also Earth to get out to Jupiter, and then on to Pluto. This enabled a smaller rocket to do the job of getting to Pluto, but it meant the Pluto 350 craft would be in space for fifteen years en route to Pluto and would have to take the heat near hot Venus before the long journey out to cold Pluto. It was not the ideal way to go, but it kept costs within the tight mission box the study team was aiming for, and the sheer ingenuity of it turned heads.

The mission plan was featured that fall in a widely read article by Farquhar and Alan called "Pushing Back the Frontier: A Mission to the Pluto-Charon System," published in *Planetary Report*, the lavishly illustrated and slickly produced newsletter of The Planetary Society, an organization with tens of thousands of members, started by Carl Sagan largely to galvanize public support for increased planetary exploration. The article began, "In the past three decades humanity has sent spacecraft to all the planets in our system except Pluto," and ended, "Whether humankind is willing to devote the resources to explore this fascinating pair of worlds [Pluto and Charon] is unknown—it is something we must decide."

The piece made a compelling case for why we must explore Pluto, why the time was right, and how the Pluto 350 study proved it could be done with a small and low-cost mission. Alan and Farquhar hoped that by placing the article before the *Planetary Report* readership they could help generate both public interest and Planetary Society lobbying support for a Pluto mission.

Around the time their *Planetary Report* article was published, a NASA press conference was held to announce the Pluto 350 study results. Alan and Fran and several others were invited to give talks about the scientific potential. The event was unusually well attended, and NASA began to notice that, as at AGU and DPS, Pluto always draws a crowd.

In front of the microphones, and in the glare of flashbulbs and television lights, members of the Pluto Underground looked at each other in disbelief at all the fuss. Goal number one had been accomplished. They were aboveground now, and with a proof-of-concept mission design. But goal number two was more daunting: Could they convince NASA to start a more serious and comprehensive mission study that might actually turn into funding for a mission?

3

TEN YEARS IN THE WILDERNESS

A NEW START?

Onward to the outer solar system! After the success and the buzz of the Pluto 350 study, the Plutophiles felt the wind at their backs, pushing them outward toward the uncharted trans-Neptunian depths, the terra incognita of our planetary system. For a moment it seemed like smooth sailing.

But their dreams and machines could not leave Earth without first navigating a perilous terrain that these young scientists were scarcely prepared for, or familiar with. Before any interplanetary trip could be launched, they first had to navigate "the beltway"—the complex political landscape surrounding Washington, DC—where the money is allocated for deep-space exploration. They had to traverse through and triumph over that DC funding swamp, and this part of their quest would ultimately prove longer, and was in some ways more arduous, than the journey from Earth to Pluto itself.

Shortly after the Pluto 350 press conference, Alan and Fran went back to NASA Headquarters and sat down with Geoff Briggs and other NASA officials to address what the next steps should be to turn it into a real mission. In the lingo of the trade, what they ultimately needed was an approved project, called a "new start."

Mission concepts are always being conceived, ranked, and studied. But what really counts is when NASA commits to a mission by putting it in its budget and proposing it to Congress. Only then—when the funding to design and build it is allocated—does a project achieve a new start. The officials at NASA told Alan and Fran that to warrant a new start, they would need the endorsement of the Solar System Exploration Subcommittee, the SSES.

HIGH COUNCIL

Within the universe of NASA mission politics then, there were a few key advisory committees that were crucial in getting a green light for any new project. Even the best plans fail if they do not pass these hurdles. In the 1990s and early 2000s the most influential group advising NASA on planetary mission strategy and new starts was the SSES; to go anywhere in the solar system, a mission had to get their approval.

It wasn't quite like the Jedi High Council, where Yoda and the wise elders meet in a grand chamber with magnificently sloping windows looking over the planet of Coruscant, deciding matters of galactic importance—but for NASA planetary missions back then, the SSES was as close as it got. The dozen or so NASA-appointed advisors of the SSES usually met in comparatively bland, windowless, rectangular conference rooms at NASA Headquarters in Washington, DC. These were not all-knowing masters, but scientists taking time out of other professional duties to make recommendations on planetary exploration strategy. And there were always many more good ideas than available funding.

Mostly, the SSES committee members listened to reports detailing what various missions could contribute to the field, and how much each would cost, then they ranked and debated priorities, and prepared "roadmap documents."

In late February 1991, the SSES was asked to pass judgment on the concept of a Pluto mission. They had received the Pluto 350 report and a document written by Alan, Fran, and colleagues describing the detailed scientific rationale for exploring Pluto and presenting a list of scientific questions that Pluto 350 could resolve. A sampling of those questions included:

> How does Pluto compare with Neptune's planet-size moon, Triton? Are they really twins left over from an early massive population of icy dwarf planets?
>
> Is Pluto's surface composition as varied as its surface markings seem to indicate? Is it really made out of completely different materials in different areas?
>
> How deep and mobile are Pluto's volatile ices? Are they merely a thin coating plated on the surface, or do they really form a deep, icy crust?
>
> Could Pluto be internally active?
>
> How does the geology of Pluto's moon, Charon, compare with Pluto's?
>
> What is the structure of Pluto's atmosphere? How quickly is it escaping from the planet?
>
> How do Pluto's strong seasons affect the surface and atmosphere? Can seasonal effects explain the high surface contrasts and the visible differences between the northern and southern polar caps? Can they explain why Pluto is darkest on the side directly facing Charon and brightest on the opposite hemisphere?
>
> What is the origin of the Pluto-Charon binary? Did the binary require a massive impact event to form it, like the one that created the Earth-Moon system?

In their reports to the SSES, the Plutophiles demonstrated convincingly that a well-equipped flyby mission could revolutionize knowledge of Pluto, addressing all these compelling questions, and more. Their case had been almost two years in the making; it was solid and could

not be ignored. The chair of the SSES at that time was Jonathan Lunine, a young and impressively accomplished, brashly confident, and
widely respected professor from the University of Arizona, who was
himself a barely closeted Plutophile. Yet some on the SSES felt that a
Pluto mission was not ready for prime time, that the idea had risen
too quickly from obscurity to deserve consideration alongside other
concepts that had long been under development. Others worried that
it would take far too many years to accomplish and that with limited
resources NASA should focus on projects promising faster payoffs—
meaning shorter flight times to closer locales.

Fortunately, strong and influential voices got behind the idea of a
Pluto mission. Geoff Briggs, who, as chief of Solar System Exploration
at NASA, had given Alan such an important foot in the door, had
stepped down. His successor, Wes Huntress, was an accomplished
planetary astrochemist who had worked for many years studying
planetary atmospheres at JPL before being promoted to NASA Headquarters to run the Agency's planetary program. At the February 1991
SSES meeting, Huntress argued that given the obvious pairing of
scientific and public interest, Pluto should be among NASA's highest
priorities for a new start.

But despite Huntress's support, a debate erupted at the SSES between the Pluto supporters, who were mostly younger scientists, and
Pluto detractors, mostly older scientists. Alan recalls that in that
debate, one key older voice stood out for Pluto, that of sixty-eight-
year-old Donald Hunten. Among planetary scientists, Hunten—an
atmospheric physicist—was a living legend, responsible for much of
the mathematical machinery used to describe and understand the work-
ings of planetary atmospheres. A reserved, no-nonsense Canadian, he
was an imposing presence. His voice was a low, gravelly growl with
only two volume settings: barely audible most of the time, or very
loud when he was riled. Though he could be intimidating, Hunten had
a reputation for rigor and fairness, and he was one of those people
whose opinion everyone valued because he had impeccable scientific
intuition: simply put, Hunten knew so damn much, and he seemed to
always be right.

At that SSES meeting, Hunten got up and spoke at a crucial time in the debate, after Alan had come under attack for pitching the Pluto mission for the next new start. Someone argued that Pluto could wait because Mars was more important and easier to reach, so Hunten stood up, eyed the room, and summarized all the compelling scientific reasons for a mission to Pluto. Then he declared in his louder, shouting voice, "God damn it! I don't expect to be alive when a mission arrives at Pluto, and if I am alive, I don't expect to be aware of the event. But this is what we should be doing. The science is important. Let's get on with it."

Hunten's remarks and gravitas turned the tide. At the end of their meeting, the SSES produced an influential report ranking a Pluto flyby in the highest priority category for new missions in the 1990s. That didn't guarantee a new start, but it meant the idea had successfully made the transition from upstart to serious contender, that it would be among the highest ranked candidates to be considered by NASA for funding.

Things were lining up. The Plutophiles had done what they needed to do by garnering the support of the SSES. As a result, Huntress chartered a new, high-level scientific advisory committee called the Outer Planets Science Working Group (OPSWG) to shepherd the Pluto mission concept through its next development steps, and he appointed Alan to be its chairman.

THE HOUSEBOAT

As we mentioned, in the early 1990s there was an effort within NASA to move away from rare, once-a-decade, multibillion-dollar "do everything" missions carrying elaborate packages of scientific instruments, toward more frequent launches of smaller, less expensive, more-modest, and more-focused missions. The modest and focused Pluto 350 mission was primed to take advantage of that.

But there was—at the same time—another movement in planetary exploration circles having the opposite effect. Some mission designers and NASA managers pointed out that each new planetary mission seemed to reinvent the wheel, designing new spacecraft from scratch. What, this faction asked, if they could develop a standardized spacecraft that would be outfitted—with customized instruments and components—for many different planetary destinations? Wouldn't such a standardized spacecraft be a way to save money on each flight and thereby enable more missions? This noble goal became embodied in the Mariner Mark II concept.

Wes Huntress was sympathetic to the move toward smaller missions. But at the same time he saw that the Mariner Mark II concept had a lot of momentum and support, particularly at JPL—the most experienced of NASA's planetary mission development centers, where Wes had worked long before coming to NASA Headquarters.

So almost immediately after the formation of OPSWG, he directed Alan to study a much larger Pluto mission than Pluto 350, one that would share its spacecraft design with both the upcoming, giant Cassini Mariner Mark II orbiter to Saturn, and another large NASA mission to orbit a comet using a Mariner Mark II.

Because Cassini and the comet orbiter were planned to be huge missions with massive complements of instruments, molding a Pluto mission around them was the opposite of the small, focused mission the Plutophiles had been arguing for. Huntress was essentially asking Stern and OPSWG to abandon the lean design of Pluto 350 for a giant "Christmas tree" mission to Pluto, weighing more than ten times as much, carrying a much more extensive load of instruments, needing a much larger rocket, and carrying a vastly higher price tag.

Alan didn't like it, but Huntress was in effect his boss in this, and so he complied. "I thought it was just crazy. I thought we'd be lucky to get our simple little, inexpensive Pluto 350 to be funded. How was the larger planetary science community going to sign up for, or NASA afford, this much more expensive 'houseboat' mission to Pluto?"

When the Mariner Mark II Pluto study was completed in late 1991, OPSWG found that it would cost more than $2 billion. Realizing that was not affordable, OPSWG strongly recommended that NASA pursue something more along the lines of the leaner Pluto 350. By early 1992, with other budget problems on his desk, Huntress agreed and relented on the push to explore Pluto using Mariner Mark II. The young Plutophiles breathed a sigh of relief. The Mariner Mark II detour had been avoided. With that settled and with the SSES's high ranking for Pluto 350 from the previous year, they hoped the way forward was now clear, and that they were on their way to starting the project.

But little did they know that out in California, there was a fly that was about to get seriously stuck in their ointment.

THE HAMSTER

Following the completion of the Voyager mission's exploration of the giant planets, in October 1991 the United States Postal Service issued a set of nine postage stamps celebrating the many successes of American planetary exploration. The set included a stamp for every planet, each illustrated with a picture of the planet and a citation to the first space mission to explore that world. But for Pluto—the only planet not then visited by a spacecraft—it was illustrated with a vague and bland artist's guess, and text that simply read PLUTO NOT YET EXPLORED.

The stamp set was released in a first-day-of-issue ceremony at JPL. A couple of young spacecraft engineers at JPL saw the NOT YET EXPLORED Pluto stamp and took it as a challenge. They asked, "Why not explore Pluto?" One of those bright young engineers was Rob Staehle, a project manager and a bit of a nonconformist. The other was Stacy Weinstein, a mission designer steeped in orbital mechanics—a talented engineer who had already worked on several successful planetary missions. Together they decided to take Pluto's unexplored status as a personal challenge to overcome. Unaware of the past two years of work going on in the scientific community to get a mission to Pluto, Staehle and Weinstein took that stamp set to their boss, Charles Elachi, then the head of planetary exploration at JPL, and pitched a radical Pluto mission study.

Staehle and Weinstein wanted to explore the possibility of sending a truly *minuscule* spacecraft to Pluto. They set their target mass at thirty-five kilograms. For comparison, this was just one-tenth the mass of Pluto 350, which was already a very small spacecraft. They planned to use new miniaturization technology, some of it borrowed from Defense Department projects that Staehle had worked on, to design a tiny spacecraft for a tiny planet. They reasoned that such a lightweight craft could be accelerated to extremely high velocities using available rockets, and thus could reach Pluto very quickly. As opposed to Pluto 350's circuitous route, which planned to use Venus, Earth, and Jupiter flybys to gain energy over a flight of almost fifteen years before it could fly on to Pluto, they would be able to send their bird directly to Pluto, getting there in just half the time. They called this mission concept "Pluto Fast Flyby," and they convinced Elachi that it was worth looking into.

Elachi provided Staehle and Weinstein with the funds to put together a first-cut design. It was very simple, and had very little capability—only carrying two scientific instruments.

When OPSWG heard this plan, they didn't like it. They didn't think it could really be done as inexpensively as Staehle and Weinstein claimed, or as quickly as they promised, but most critically, they thought the mission went too far in skimping on scientific return. In contrast to the Mariner Mark II "houseboat" they had successfully argued against sending to Pluto, now they were arguing against sending a hamster to Pluto. Pluto 350, OPSWG argued to NASA Headquarters, was the sweet spot between a houseboat and a hamster, and should be started without further delay.

THE HOLLYWOOD MANEUVER

What happened next was as unlikely as anything cooked up in a Hollywood script. In fact, it took place at a ceremony in Beverly Hills, in an ornate auditorium at the headquarters of the Academy of Motion Picture Arts and Sciences. Rob Staehle was upset that OPSWG had rejected his plan. Ironically, the Pluto Underground had become, in his eyes, the conservative establishment. Staehle thought they were rejecting a chance at genuine innovation in favor of more-tried-and-true methods. He also knew that there was a new alignment of power coming to NASA that would be sympathetic to his approach.

That new power alignment: on April 1, 1992, Dan Goldin began work as the new NASA administrator, appointed by President George H. W. Bush. As we previously mentioned, NASA at this time was encouraging the development of smaller, less ambitious planetary missions, with more-modest and -focused goals, containing fewer instruments but allowing launches to occur more frequently than in the past.

Goldin was an aerospace executive who wanted to shake up the culture of NASA, which he thought was too attached to giant, expensive spacecraft. In addition to pushing for smaller missions, Goldin also wanted to encourage more risk-taking. By his logic, if NASA flew a larger number of smaller missions, then they could individually be less risk-averse; that is, if NASA lost the occasional mission, it would be less of a setback because there would be many others. Goldin also reasoned that NASA could save money by taking more risks, cutting back on the tradition of rigorous testing and its conservative, cautious approach toward deploying newer, untested technology. Thus was born Goldin's mantra, which became NASA's then guiding philosophy: "Faster, Better, Cheaper," or FBC, as it became known.

Wes Huntress recalls that Goldin, his new boss, wasted no time in explaining the new guiding philosophy to him, and that Goldin seemed to exhibit a certain naïveté about what was realistically possible. Huntress:

> When I was introduced to Dan he looked me hard in the eyeballs, poked my chest, and said, "Aha, you're the planetary exploration guy. I want you to send a mission to Pluto to get a sample from the surface and return it to Earth in less than a decade and do it for less than $100 million." I was so shocked that I blurted out something like, "Well, that certainly is a challenge. We'll have to take a look at it." I wanted to tell him that it was simply

not possible. The fact that I didn't probably saved my job considering the number of Associate Administrators and other top NASA employees Goldin fired over his first year.

Rob Staehle felt that if he could only get Goldin's attention, the new administrator would be sure to embrace his Pluto Fast Flyby idea. Rob knew he needed to find a way to circumvent official channels and get his mission concept right in front of Goldin. He found that chance when a friend who worked as an usher at the Motion Picture Academy theater told him that Goldin would be attending a ceremony at the Academy in L.A., near JPL. Rob recalls:

> She called me up and said, "Rob, there's a thing happening here at the academy that you might be interested in. It involves your new boss." I said, "What do you mean my new boss?" She said, "Well, the new NASA administrator, Dan Goldin. You know who he is?" I said, "Yeah, I know who he is. But I've never met him, and I know hardly anything about him." She said, "Well, it turns out that when he arrived in the NASA administrator's office and he opened one of his desk drawers, there was a gold Oscar statue in the drawer, with a note from his predecessor, Dick Truly, saying, 'I didn't get an opportunity to return this to the Motion Picture Academy; maybe you could take care of it.'"

The statue had been flown on the space shuttle during the Academy Awards festivities earlier that year. The shuttle crew had participated in the ceremony, replete with a weightless, floating Oscar statue, as Steven Spielberg presented a lifetime achievement award to George Lucas. Now Goldin, along with a few shuttle astronauts, was coming to return it to the Academy. Rob's friend landed him an invitation to the ceremony.

Staehle approached Goldin after the ceremony. Surrounded by Hollywood types and a smattering of NASA people, he introduced himself and said, in effect, "Mr. Administrator, I'm a JPL engineer in charge of a Pluto mission study. We have a breakthrough way to do this at low cost with revolutionary technology, but the establishment won't let us do it. I can get us to Pluto by the late 1990s with a very small spacecraft. The package I am holding is a study that proves it. Can you help?" Goldin said, "Can I have that?" So Staehle handed him his report with all the mission details. Goldin promised he would read it later that evening.

Goldin then quickly embraced the Pluto Fast Flyby concept. Upon his return to Washington, he told Wes Huntress, "I want you to do this." Wes, again knowing his new boss's affinity for firing naysayers, contacted Alan and told him that OPSWG had to drop Pluto 350. Instead, NASA was going to pursue Staehle's Pluto Fast Flyby mission. "That's what the administrator wants," Huntress said, "and that's what we're going to do." Alan knew immediately that this was bad news:

> I just thought to myself, "We are so screwed, because this thing is not going to work." You could just see that it involved too many development miracles that would either cause it to become too expensive, or it would end up growing in mass and not really be able to travel as fast as Rob's team had promised Goldin, or that the SSES was going to see that it was too light on capability and walk away from it. The end result: I thought we would likely spend a year or more trying to develop this and nothing would come of it. And you know, that's exactly what happened.

At Goldin's direction, Huntress provided Staehle with more funds to further flesh out the Pluto Fast Flyby concept. But in less than a year, Staehle's own team proved that the idea could not work as initially conceived in their earlier JPL study. They could not do a 35-kilogram mission. Even a stripped-down, bare-bones spacecraft with only two instruments—a camera and a radio-science experiment to probe Pluto's atmosphere—came out at more than 100 kilograms. And this was without any backup systems to make the spacecraft reliable enough to undertake the nearly decade-long mission.

Countering the fatal criticism that having no backup systems was too risky for a mission so long, Staehle's team then produced a more robust version, but this, of course, added weight, bringing their probe now to 164 kilograms—half of Pluto 350's mass, but with far less capability. Cost estimates had also steadily risen, from $400 million to more than $1 billion—rivaling Pluto 350's cost.

Staehle's team brought a full-size mock-up of its spacecraft to the 1992 World Space Congress in Washington, DC. The team thought it had really accomplished something, but when Dan Goldin heard about the dramatically higher mass and cost, he became incensed. "What happened to 35 kilograms?" he asked.

Perhaps Goldin felt he had been sold a bill of goods, and that his so-called beautiful dream of an inexpensive, head-turning Pluto Fast Flyby had gone up in smoke. Staehle's concept had also become a nightmare for the Pluto community, which had been directed to study only that concept, with no alternative.

Alan surmises now that it was at this point that Goldin decided that the Pluto community (which for him was embodied by Rob Staehle) had been disingenuous in their promises. Both Goldin and Huntress maintain today that they remained committed to flying a Pluto mission as soon as possible, but Alan believes this was the moment Goldin soured on it.

The end result: work toward Pluto missions did not stop, but the road became rockier. Goldin still publicly spoke of Pluto as a high priority, but now new demands and hurdles kept appearing from his office. What was to follow were frustrating years of seemingly endless mission studies, cancellations, then new mission studies, and then new mission cancellations.

Around this time two other big obstacles arose to make matters worse, both were crashes: first a budget crash, then a rocket crash. The first was the release that February of the president's budget for 1994, in which money for planetary missions was flatlined by the White House. An expected increase, which Huntress had been counting on to fund the start of new missions to the outer solar system, had been entirely erased. The second big setback came just months later, in August 1993, when NASA's Mars Observer spacecraft blew up three days before it was to fire its engine to go into orbit around the red planet. This orbiter had been conceived as NASA's triumphant return to Mars, ending a long hiatus since the Vikings, which were launched in 1975, but now it had become space junk orbiting the Sun.

Goldin's response to the Mars Observer failure was typically bold: he started an entirely new program of multiple spacecraft to be sent to Mars, replacing the lost science and doing much more, with a whole series of missions to be launched over many years. He would use this new Mars program to implement his "faster, better, cheaper" philosophy, investing less in testing and redundancy, and accepting more risk as a trade-off against more-frequent missions. To fund the new program Goldin swept up all the money he could find. As Alan recalls:

> Goldin told us, "I love Pluto, but I've got a new Mars program to fund and you've got to get it down in the range of $400 million including launch costs." That just seemed like an impossible assignment. Pluto 350 had a much higher price tag, and even Staehle's stripped-down idea for a mission was over $1 billion. Launch costs alone those days were nearly $400 million. Goldin's new dictum all but foreclosed any way of doing a mission to Pluto, unless we could find a clever way out of his cost box.

THE RUSSIAN GAMBIT

Stern's OPSWG and Staehle's JPL Pluto mission office were still joined at the hip by NASA headquarters, and they were in lockstep on one point—that Goldin's new cost target just didn't make sense. The Voyager mission had cost nearly ten times that. How could anyone do a Pluto mission for just $400 million? The high prices of launch vehicles alone seemed to make a trip to Pluto impossible under these circumstances, unless they found an overseas partner to pay for the launch.

At that time, in the economic disruption that came in the wake of the Cold War, the former Soviet space program was moribund.

Their planetary exploration program, which in past decades had been so successful at landing spacecraft on the Moon and on Venus and intercepting Halley's Comet, was barely on life support. They had capable scientists, and they had huge, reliable rockets, called Protons. But they had no interplanetary spacecraft to launch on these rockets, and no resources to build new spacecraft. They also lacked NASA's experience with missions to the outer planets.

In Alan's view, the United States and Russia each had what the other needed. So Alan envisioned a joint project: Russia would provide their powerful Proton launch vehicle, America would build and fly the spacecraft, and both would share the glory. In the then-thawing relations between the two countries, Alan thought the ticket to Pluto could be a triumphant closing chapter to the first exploration of the planets: a joint mission by former rivals to the last and the farthest of all the known planets.

Feeling fed up with the impossible request to fly a Pluto mission for under $400 million, and excited by the idea of getting a free launch, Alan therefore decided to change the game. With Staehle's knowledge, but otherwise more or less on his own, Alan decided to reach out to Russian counterparts to explore the possibility of getting them involved. Without anyone's permission, he flew to Moscow and went to see Alec Galeev, the head of the prestigious and pioneering Space Research Institute in Moscow, the Russian equivalent of JPL. Alan:

> I'd never met the man, but I knew he was very powerful and he was our hope for getting a Pluto mission done within Goldin's cost box. So, I pitched him, saying something like, "Your country has never been able to do the outer planets. We'll make you part of our very highly visible Pluto mission, the Everest of planetary exploration. What we want from you is a contribution of the launch vehicles. It will help resurrect your planetary program, and we'll put Russian engineers on the project. We'll teach you everything about outer planets exploration, and you will get the credit and pride of launching the first mission to the last planet."

Huntress got wind of this plan from JPL just before Alan left on the trip and tried to stop him. Literally, when he was at the airport leaving. Alan:

> I got a phone call saying, "Don't get on the airplane! No one's given you permission to go talk to the Russians!"

Wes:

> I think what Alan did was irritate some of the folks in the international office at NASA Headquarters that we had this rogue scientist going over there to talk to our main contact for planetary cooperation. What's this guy doing? He's not part of our protocol. It kind of ticked them off. I just had a cup of coffee and kind of giggled.

But Alan knew there was no way to meet Goldin's price target without free launchers, and playing by the rules had been getting them nowhere. Alan recalls:

> So, I got on the airplane, I went over there, I pitched it, but Galeev and his folks told me, "Nyet." He said there wasn't enough in it for them and they didn't want to just be used for the launch on day one, with everything else in the mission being American. I remember it was January 1994, and it was blizzarding hard outside in Moscow. Just horrible weather. It felt as cold as Pluto.
>
> But I went back the next day and Galeev said, "We have a new proposal. Your American spacecraft could also carry a Russian probe that would separate and enter Pluto's atmosphere with a mass spectrometer just before it would crash-land on Pluto." This way the Russians would have a really unique part of the exploration at Pluto, and they could boast that they, not we, had actually touched Pluto. Galeev said, "If we can do that, I can get you the Proton launchers." I told him I thought I could

sell this back home, so we adjourned to a banquet over vodkas and Georgian red wines.

The Russian atmospheric-probe idea made the trip a success. I was so excited that I didn't even fly home to Colorado; I went directly to JPL, sat down with Staehle and Weinstein and their whole team. I explained, "This is how we can make it work: it's the end of the Cold War. It's the Russians and the Americans doing the last planet together." They liked the concept too, and they also felt it was a victory, as a way to get in Goldin's impossible cost box.

Alan next took the idea to Huntress at NASA Headquarters. Though he was annoyed at Alan's rogue diplomacy, Huntress was actually enthusiastic about the idea of working with the Russians on this. So Huntress sold the idea to Goldin and organized a NASA trip to the Russian Academy of Sciences to explore the joint mission concept further. With Alan and other Pluto scientists in attendance in Moscow a few months later, Huntress pitched the Pluto mission as "The first outer planet mission for Russia, and to the Siberia of the solar system as well!" It was a long shot that introduced a whole new set of unpredictable variables involving international relations and diplomacy into the already complex game of getting a Pluto mission funded and approved. Yet Huntress felt it was worth a try.

But no sooner did Huntress come aboard than the Russians decided they would not provide a launch vehicle for free. The price they instead demanded was substantially cheaper than using an American rocket, but not zero.

At that time, U.S. government agencies like NASA were, by law, not allowed to purchase Russian rockets. So the Americans needed to involve a third country to pay for the Russian Proton rockets. Alan started asking around and learned that the German space agency might be interested, if their scientists could add their own probe to be dropped off during the planned Jupiter gravity assist flyby, to study Jupiter's moon, Io.

But it got even more complicated on the American side. There was a new concern within NASA against launching the mission's must-have nuclear power source on a Russian launch vehicle. Many felt that this would never be approved by other U.S. agencies that had signature authority on nuclear-powered spacecraft—like the Departments of Defense and Energy and the EPA.

It all became a bridge too far. The Russian launcher gambit that had looked so promising soon disintegrated. Now, 1996, they were back to square one.

A DREAM DEFERRED

For the Plutophiles, the late 1990s was a low point in morale. They had been at it for over half a dozen years, week in and week out. They had felt so close to their dream in 1991–1992, only to see it devolve into a maze of frustrations in the years since. Having once seemed likely to be fast-tracked at the beginning of Dan Goldin's tenure, a mission to Pluto was no longer the shiny new thing; it had become battered and worn.

In fact, now there was another new darling of the outer solar system: Jupiter's moon Europa, which had been revealed by the Galileo Jupiter orbiter to likely have an ocean inside it—something then unprecedented off Earth. Many in the planetary community saw Goldin's interest in that and started to feel that their new highest priority should be to send an orbiter to Europa, following the trail of possible life in that vast ocean.

The Pluto supporters were frustrated. Every time Alan's OPSWG and Staehle's Pluto team fit the mission in Goldin's box, the rules would change. Several times they got through the SSES and through Huntress, who by then headed all of space science, only to learn that Administrator Goldin wanted to give them yet another new and frustrating assignment. Alan:

> It took me years to figure out the pattern, because Goldin always told us we were one of his favorites, and he always funded about a $30-million-a-year budget line item to do Pluto technology

development and mission design studies. But he never let us actually get to the head of the line and get a new start to build a mission. Every time we got close, he had some new reason to send us back to the showers with more studies.

At one point Goldin even told us, "I'm all for this, but you have to figure out how to do it without nuclear power." We were incredulous: "What? Are you kidding? How can anyone run a spacecraft so far from the Sun without nuclear power?" Maybe I was a little paranoid, but this was the point at which I started to suspect Goldin was just toying with us, that he was never going to start the mission; because there was always another reason to delay, always another reason to study some other concept. So by late 1996 I began to suspect that Goldin just kept sending us on errands, and I wasn't alone. Errand after errand after errand, each taking six months or a year to resolve. It seemed he was never going to let us out of the study box, but we knew we couldn't quit: because if we ever walked away he'd drop the project altogether, and there would be no hope to get a new Pluto mission started. So we decided to wait him out.

THE RISE OF THE "3RD ZONE"

For decades, planetary scientists suspected that Pluto might not be all alone in the solar system's outer reaches. Indeed, even as far back as the 1930s and 1940s, searches for more bodies in Pluto's distant region were conducted—but nothing was found. More searches took place in later decades, but again they came up empty. Nonetheless, the idea that Pluto wasn't alone gained traction from time to time as various planetary scientists made mathematical and other arguments supporting the possibility of a cohort population orbiting with Pluto.

By the time Voyager 2 was exploring Uranus and Neptune in the late 1980s, Bill McKinnon had become a key proponent of this idea. His logic was based on his analysis of Neptune's moon Triton, so similar in size to Pluto, and his deduction based on Triton's orbit that Neptune had captured it from a former Pluto-like orbit around the

Sun. If there had been two such bodies, McKinnon reasoned, why couldn't there be more? Perhaps even a lot more.

The possibility that there was a vast swarm of such objects had been earlier suggested by pioneering mid-twentieth-century planetary science giant Gerard Kuiper, who sought to explain the origin of the planets. In 1950 Kuiper had proposed that the process of accretion (the building of larger objects from smaller ones), which originally formed the Earth and other planets, should have left behind a huge number of "planetesimals," or small building blocks of planets, out beyond the outermost of giant planets, Neptune.

Perhaps, McKinnon reasoned, as Kuiper and others had, Pluto was not just an oddball outlier but our first glimpse of what would later come to be known as the Kuiper Belt.

In 1991, Alan took this idea further, publishing a research paper called "On the Number of Planets in the Outer Solar System: Evidence of a Substantial Population of 1000-km Bodies," arguing mathematically, from several kinds of forensic evidence in the outer solar system, that there should be hundreds or perhaps thousands of small planets that had been created early on, constituting a whole new "3rd zone" of the solar system, beyond Neptune.

For over a century we had referred to the zone between Jupiter and Neptune, the realm that the Voyagers had explored, as the "outer solar system." Alan's 1991 paper now claimed that the giant planets region might actually be the middle zone between the inner solar system, where Earth orbits, and the "real outer solar system," where Pluto and its vast cohort lie.

By the time of Voyager 2's Neptune encounter, Earthbound telescopes and detectors had improved to the point that such "Kuiper Belt Objects" (KBOs), if they were indeed there, could in principle be detected. It took a few years of searching, but the new tools of electronic imagers (CCD cameras, which are in all our cell phones today) and massive computer searches (that removed the painstaking workload Tombaugh had endured) made all the difference. Beginning in 1992, planetary scientists began to find KBOs—and thus Pluto's cohort!

At first it was just a handful, but then more and more were discovered. Ultimately in the 1990s, more than a thousand KBOs were
found, distributed in the wide zone beyond Neptune that came to be
called the Kuiper Belt. Most were small, perhaps the size of just coun-
ties, but others were much larger, and a handful even rivaled Pluto
in size—the size of continents. Today, based on what has been found
and the vast areas still not mapped for lack of labor and funding, there
are estimated to be more than one hundred thousand objects greater
than fifty miles in diameter there, and an even larger, unknown number of still smaller ones.

Fran Bagenal recollects how these discoveries added fuel to the fire
for a Pluto mission:

> Interest just exploded once we started to see the Kuiper Belt
> Objects. It was as if it had just been a bunch of Pluto people who
> were interested in this lone, fascinating object at the edge of the
> solar system and could make a strong case for its exploration. Then
> it suddenly became "This is a whole new frontier!" I remember
> sitting in my office at CU Boulder where we were meeting to
> write a document to make the case for a Pluto mission. Bill
> McKinnon was there, and he pointed out the window to the Rocky
> Mountains and said, "This is like the new lands to the west that
> have not been explored. We should go and explore them. What's
> out there?" I realized that he was right, and it got me really fired up.

The discovery of the Kuiper Belt provided the additional scientific
motivation to raise the priority of Pluto exploration back to the top.
To emphasize that their hoped-for mission was now not just about
exploring Pluto, but about exploring this whole third region in the
solar system, the mission to explore Pluto was renamed "Pluto Kuiper
Express" (PKE).

PKE was the fifth effort to get a Pluto mission, which had begun as
Pluto 350 and had been recast and renamed again and again to adapt
and try to find just the key that would unlock NASA funding. Never

lacking in persistence, the Plutophiles energetically went forward to define this new JPL concept.

To put meat on the bones of PKE, and to prepare for issuing a call for the mission's scientific instrument proposals, Huntress formed a Science Definition Team (SDT) of Pluto and Kuiper Belt experts. The SDT was charged with defining formal goals for a Pluto mission and basic specifications for the instruments it would carry to achieve these goals. It was a very good sign.

Forming the SDT signaled that the agency had again become serious about a Pluto mission, because NASA used SDTs to kick off formal planning to create a mission new start. Planetary scientist Jonathan Lunine, who had earlier chaired the SSES, was asked by Huntress to chair the SDT. Lunine's team included Alan and many old members of the Pluto Underground and OPSWG, and also new faces who were experts in the science of the Kuiper Belt. Lunine's SDT worked for almost a year to craft a tight mission rationale, a list of necessary and nice-to-have scientific instruments, and a detailed scientific case. Their report emerged to rave reviews from multiple sectors in the planetary science community.

As a result things were looking up again. But then, in late 1998, longtime Pluto mission champion and NASA Headquarters science head Wes Huntress left NASA. (Huntress told us in 2016 that his main regret in leaving was that he had failed to secure a new start for a Pluto flyby mission.) Unfortunately, his successor as Associate Administrator for science, an astronomer named Ed Weiler, was much less committed to that goal.

Weiler did, however, keep some funding going, and in 1999, after Lunine's SDT issued their report, Weiler's office put out a call for teams to compete via a proposal process for funding to actually build instruments for PKE. This also seemed promising: real money to build real instruments that would fly to Pluto. The way it was structured, there would be four sensors on PKE (cameras, a composition spectrometer, an atmospheric spectrometer, and radio science to probe the atmospheric temperature and pressure), but only two winners would be selected: one for a combined package with the first three of these sensors ("the remote sensing suite") and one for radio science which really was a different kind of instrument that didn't fit naturally into the other group. The proposal competition became a "shoot-out" to get aboard what looked like the one and only train leaving the station for Pluto. Winning it was do or die. Competing teams formed at various labs and universities around the country, each with a team leader which, we pointed out earlier, NASA refers to as a principal investigator (or PI).

Alan's team proposed one of the combined imager/spectrometer suites of instruments, and it included many of the original young guns from the Pluto Underground and OPSWG. Alan believed that his most formidable competition was a powerful team of Voyager veterans from JPL, led by a U.S. Geological Survey planetary geologist named Larry Soderblom, one of the most respected names in all of planetary science—then a god of the field. How could Alan and his relatively young team possibly compete with Soderblom's team's experience? Well, for one thing, Alan recruited his own god, in the form of Eugene Shoemaker. Shoemaker was regarded as the founder of the field of planetary geology and had worked on virtually every planetary mission. Shoemaker was probably—no exaggeration—the most widely respected and well-liked person in the field. Nobody had more gravitas. Alan:

> I felt like Larry had the "in." He had the better connections at NASA Headquarters, he had JPL's A-team of engineers, and he had a really phenomenal science team. I had a great science team, too, but I always felt that we were largely a group of young insurgents, compared to the master: a guy that has been on Voyager and many other projects, and it felt like they were the incumbents— like it was theirs to lose. The idea of bringing Gene Shoemaker on board was to counter that. We formed our team anticipating a NASA competition later in 1998, or in 1999. At that time Gene was sixty-nine years old, and I had to really talk him into doing something that was going to be a fifteen- or twenty-

year project, but he agreed to do it. I remember him saying to me, "I thought I was too old to be aboard Cassini Saturn orbiter proposals a decade ago, but this is just too much fun to pass up!"

Alan and his team worked tirelessly on their proposal, refining for almost eighteen months. To make it stronger, outside experts were brought in to critique every aspect of the proposal, to help improve the designs, management plans, plans for education and public outreach, schedules, and more. The other competing teams of course did the same.

Proposals were finally due in the spring of 2000. NASA then convened a suite of expert review panels to evaluate all the proposals. Through back channels in the summer of 2000 Alan got word that his team had won for the camera/spectrometer suite. That was encouraging, but then . . . there was only silence. Alan:

> We kept waiting and waiting for NASA to tell us that we had won, but then out of the blue I got a phone call from Rob Staehle in late August of 2000, and he said, "It looks like Weiler has decided to cancel the whole thing." How, when instrument selection decisions had apparently already been made and a mission new start finally seemed imminent, could this have happened? I was floored.

THE STOP-WORK ORDER

As it turned out, while the instrument proposals were being evaluated, NASA Headquarters discovered that engineers from JPL had created a budget monster. Weiler had been promised PKE could be done for less than $700 million. But when Weiler ordered a cost review, it was found that JPL's real cost was going to be nearly twice that, maybe more.

Weiler decided enough was enough. He was sick and tired of Pluto studies that never got off the drawing board because they kept blowing up on cost. So in mid-September of 2000, he announced that NASA was canceling the instrument proposal competition—stillborn—meaning that there was to be no winner at all. Everybody lost.

And not only was the instrument competition canceled, but Weiler issued a "stop-work order" on all further Pluto mission work.

As the head of all NASA science missions, Weiler's letter had the force of a legal document saying that no further NASA money could be spent on studying any Pluto mission. Alan recalls:

> It was stunning. After five separate efforts, endorsements from who knows how many NASA advisory committees, after a Science Definition Team and a call for instrument proposals, after spending probably $300 million on studies over ten years, Weiler's action threw it all in the can and offered no hope of reversal.
>
> In response, JPL just put everything in filing cabinets, and disbanded Staehle's team. The whole thing evaporated.
>
> Those of us who'd been working on it felt like we had been through a decade of hell running errands, with endless study variations for NASA Headquarters. How many iterations of this, how many committees had we been in front of, how many different planetary directors had we had at NASA, how many different everythings had we put up with? Big missions, little missions, micro-missions, Russian missions, German missions, nonnuclear missions, Pluto-only missions, Pluto-plus-Kuiper-Belt missions, and more; Weiler's move just threw it all away.
>
> It was the end. He'd killed it. It felt like we'd survived the Bataan Death March across the 1990s, and then just when we got to the finish line, when there was a promise of being released to build the instruments and start the mission, they beheaded you.
>
> Weiler's cancellation directive just took people's breath away, and it left us as if we were back in 1989 trying to start something from scratch.

And if that wasn't enough, Weiler also declared that NASA would not even consider another Pluto mission study for a decade—until the

2020s. He publicly called Pluto mission studies and any efforts to make a mission new start "Dead. Dead. Dead."

That was it: After a dizzying decade of false hopes, reversals, new renewals, and after persevering longer than might have been healthy, the Plutophiles had finally reached the end of the line. Goldin backed Weiler. There was no appeal.

It was over.

4

THE UNDEAD

KICKING THE FOOTBALL

With all of the work on Pluto missions since 1989 now swept off the table, the defeat of Weiler's once-and-for-all cancellation was breathtaking. Even more than any of the previous delays and setbacks, this one was utterly profound. NASA had declared itself out of the Pluto exploration business.

To Alan, this simply could not pass: too much was at stake, and too many people had invested too much. So he and the Pluto Underground did what they had always done: they dusted themselves off again and got to work. They began behind the scenes, but soon opened a second front and moved their efforts to front stage. They wanted to share their shock and indignation with both the public and their professional community and wanted to put the need for and excitement of Pluto exploration in the press at every turn. They generated letters to editors and newspaper op-eds (this was well before blogs existed), making their argument clear: don't throw this all away just because JPL made the mission too expensive.

And they also enlisted others to agitate on behalf of resuscitating a Pluto mission. Their rabble-rousing was effective: the press and the public widely decried the cancellation. An article in *Space Daily* reported, "This move infuriated many planetary scientists, because . . . Pluto is virtually unique in being a scientific subject that won't wait." And it quoted Lou Friedman, then the director of The Planetary Society, saying that the deciding factor in a resurrection could be that "NASA is surprised to see how popular Pluto really is."

Yet time was running out, because various terrestrial and celestial factors were lining up in ways that meant it could be "now or never" for Pluto exploration.

First, there was the launch window dictated by the relative motions of Earth, Jupiter, and Pluto. Just as Voyager's grand tour mission had been constrained to launch when the giant planets were in rare alignment, so, too, any trip to Pluto requiring a Jupiter Gravity Assist (JGA) to speed it on its way could only be made when Jupiter, on its twelve-year orbit, came around to be in line with Pluto. To reach Pluto before the 2020s, the upcoming 2002–2006 Jupiter launch window was a must-make.

Additionally, two different aspects of Pluto's motion in its 248-year orbit added further deadline pressure. First, in 1989, Pluto had reached its perihelion—its closest point to the Sun in its orbit—and ever since then it had begun slowly receding, making it a slightly harder target to reach with each passing year. Second, as Pluto receded from the Sun, it was cooling, and that held a potentially ominous threat for studying its atmosphere.

By the early 1990s it had been known that Pluto's atmosphere was primarily composed of molecular nitrogen, the same gas that makes up most of the Earth's atmosphere. But unlike Earth, Pluto's nitrogen atmosphere is created by the sublimation* of its snowy surface. In this process, the atmospheric pressure is very strongly, in fact exponentially, dependent on the surface temperature. As a result, each few degrees of surface cooling cuts the atmospheric pressure in half. So as Pluto traveled outward in its orbit, and the Sun's warming of Pluto's

---

*Sublimation is like evaporation, which occurs when a liquid is heated and becomes a gas; when an ice becomes a gas the same way, with no liquid phase in between, scientists term it sublimation.

surface naturally decreased, the atmospheric temperature was expected to decrease, and as a result the atmospheric pressure would begin to fall, steeply, perhaps to levels hundreds or thousands of times lower than its perihelion pressure. If that happened, the atmosphere would, in effect, cease to exist and couldn't be studied by *any* mission sent there once the atmospheric freeze-out occurred.

Atmospheric models were predicting that this pressure collapse might well happen sometime between 2010 and 2020, but almost certainly soon thereafter. For the Plutophiles, this placed a deadline on getting a flyby of Pluto. If a mission was going to observe the atmosphere, it had better get under way—now.

And if that wasn't enough, there was still more reason to hurry: Pluto's sharply tilted spin axis, angled 122 degrees from the plane of its orbit (Earth's tilt is just 23.5 degrees), creates drastic seasonal lighting changes across its globe. Think of how, on Earth, places that are North of the Arctic Circle and South of the Antarctic Circle enjoy both a midnight Sun and then perpetual night for months each year. Something similar but much more extreme happens on Pluto because its axis is so much more extremely tilted than Earth's.

As the 1990s wore on and Pluto followed its slow orbital arc around the Sun, a steadily larger portion of its southern hemisphere was entering into a decades-long season of perpetual darkness. As a result, any spacecraft that reached Pluto in 2015 would be able to observe only about 75 percent of the surface, with the other 25 percent enshrouded in a polar winter night. But by the early 2020s, only 60 percent would be visible, with more of the planet drifting into darkness with each successive year, and by the 2030s, only 50 percent would be visible. Translation: the longer a mission was delayed, the less of the surface of Pluto (and for that matter, of Charon, too) could be studied when the spacecraft arrived.

These factors—the need for a Jupiter gravity assist, the possibility that Pluto's atmosphere would freeze out, and the shrinking amount of Pluto and Charon that could be mapped—were all reasons to get the mission started as soon as possible.

An old and powerful ally in this battle was the SSES. In their meeting on the auspicious day of Halloween of 2000—barely a month after Weiler's PKE cancellation—the Pluto situation was at the top of the agenda.

The community of planetary scientists the SSES was charged to represent did not take kindly to the seemingly cavalier cancellation of a mission that had, through a painfully long process, risen to the top. They had seen their colleagues at universities and labs around the country working their tails off to create the science case and then competing to design the best possible set of scientific instruments to explore Pluto. Weiler's cancellation felt like more than just a setback to the Pluto crowd: The entire planetary science community was being jerked around by it, and finally started feeling what the Pluto crowd had been experiencing all those years.

Alan and Jonathan Lunine spoke to the SSES that Halloween, making the case to go to Pluto, presenting a hard-nosed scientific rationale to do the mission now, and forcefully defending a quick restart of Pluto exploration. They also described how the exploration of Pluto could be done much more simply and inexpensively than JPL had planned with PKE.

The problem for PKE was that the JPL estimated cost of the entire outer planets program, which now included missions to both Pluto and Europa, had exploded to almost $4 billion in today's dollars, and the Pluto mission alone had ballooned to $1.5 billion, possibly more. But as the SSES learned more, they became ever more skeptical of the way in which these costs had been calculated.

One member of the SSES who played a pivotal role was a venerable and senior Greek-American space scientist named Stamatios "Tom" Krimigis. Tom is tall, thin, and genteel. Speaking with a deep voice and thick Greek accent, he could easily play the role of a handsome Greek stranger in a classic Hollywood film.

Tom had been involved in planetary exploration from nearly the beginning and had built instruments for spacecraft that had traveled to every planet but Pluto, as well as to a number of asteroids and comets, and he had been one of NASA's first mission PIs back in the 1980s (creating artificial auroras to understand the origin of that phenomena). In the 1990s, Tom had also been instrumental in helping NASA develop the then-new competition structure for interplanetary missions. In this new paradigm, not just the scientific instruments, but everything—the design and construction of the entire spacecraft, the ground operations, and the scientific investigations—all of that would be competed for and then awarded to the one team with the best proposal led by a scientist: the principal investigator (PI).

Largely a response to the ballooning mission costs and extreme budget pressures of the late 1990s, which threatened to derail America's planetary exploration program, this new PI-led model represented a departure from NASA's usual procedure of assigning planetary missions to a big lab (usually JPL), running competitions only for the scientific instruments that would go aboard, and placing the entire project under the helm of a professional project manager.

Suffice to say, Krimigis had as much credibility as anyone in the business. But Tom had one more piece of relevant experience: he was also the head of the Space Science Department at the Johns Hopkins Applied Physics Laboratory (APL), which was emerging as a smaller, leaner competitor to JPL, able to produce less expensive planetary missions.

Tom remembers one moment at that crucial October 2000 SSES:

> [JPL Project Manager] John McNamee came to the committee to tell us why the Pluto mission budget had gone from $600 million to $1.5 billion, and why the Pluto spacecraft was also so heavy and couldn't be built for less, and so on. He made the mistake of passing around a circuit board part that had about an inch of aluminum on top of it, and he said, "Take a look at this. Look how heavy it is. We are going to fly by Jupiter on the way to Pluto, and so it needs all this shielding to protect it from the radiation. That's why we are doing this."
>
> That circuit board was circulated around the table, and when it came to me I said, "Now, wait a minute. This Pluto spacecraft

will fly by much farther out from Jupiter than Voyager did, which didn't need any shielding. This is ridiculous."

As it turned out, JPL, following a dictum from NASA Headquarters, had put expensive Europa-mission Jupiter radiation-protection design requirements on PKE, like a boat anchor, and that anchor had sunk Pluto by driving its cost skyward.

Once the SSES discovered this, they saw that the Pluto mission could be done much less expensively. So, led by chairman Mike Drake, then head of the Lunar and Planetary Laboratory at the University of Arizona—the largest planetary research institution in the United States—the SSES wrote a letter telling Weiler they did not approve of the cancellation and saw it in a larger context: as a bad sign for the health and future of American planetary exploration. The SSES then recommended that Weiler resurrect the Pluto mission, but to control costs, do it in the style of a competed mission, rather than giving it directly to JPL, as NASA had done so many times before. They further recommended that the Europa mission be delayed, if necessary, to insure adequate funding for a Pluto mission, reiterating all of the reasons why Pluto could not wait, none of which applied to Europa.

At the same time, two other significant events were taking shape to resurrect the mission. First, with Alan's cajoling, The Planetary Society organized another old-school letter-writing campaign among their members, who were fanatical about planetary exploration and who had long been excited for Pluto exploration. The Society flooded NASA Headquarters with thousands of letters protesting the cancellation. At the same time, unbeknownst to Alan, a high school student named Ted Nichols appeared on the scene. Possessing an intense passion for exploring Pluto—as well as some great PR skills—Ted attracted much attention both with his "Save-the-Pluto-mission" website and some other ambitious PR. Alan remembers:

Ted was a seventeen-year-old, and a really kind of cute-looking kid, and he just couldn't stand that the effort to explore Pluto had been canceled. Ted lived in Pennsylvania, which isn't all that far from DC, so he went on his own to NASA Headquarters in Washington and pleaded for Pluto exploration, and he brought the press with him. He was a brilliant tactician, because he made himself the poster child for, and the face of, public disappointment about NASA's cancellation of PKE. And somehow he got all the way up to Weiler's office. I don't know how he talked his way in there. NASA actually thought I put him up to it or that The Planetary Society did, but he did it entirely on his own. I didn't even know him then. And what did Weiler's folks do when the kid arrived there? They put the seventeen-year-old in a room with six adults, six NASA bureaucrats, and they began questioning him: "Who put you up to this? Why are you coming here out of nowhere? Who's backing you? Who paid for your trip?" And the kid responded saying something like, "It's just me. I want to see Pluto explored, and you have dashed my dream. How could you?"

Nichols himself and that dream of his became a media story that personalized the cancellation, embodying the disappointment of youth. And in addition, by the late fall of 2000, The Planetary Society had generated more than ten thousand letters to NASA and Capitol Hill from citizens concerned about the cancellation. Lou Friedman, then The Planetary Society's executive director, who had cofounded the society with Carl Sagan, bundled them all up, got on a plane from California, and delivered them ceremoniously to Capitol Hill . . . with the press in tow, of course. The news release read: "The American people are pleading for Pluto!"

And the barrage continued. The DPS, the largest and most influential professional group of planetary scientists in the world, at the urging of members like Stern and Lunine, issued a press release pointing out that if a Pluto mission was not started soon, the crucial Jupiter launch window would be missed, and that if that happened, there would likely be no atmosphere to study when some future probe arrived in the then crazy, far-off 2020s.

The press was beginning to pick up on these messages, creating magazine, newspaper, and even television news stories, further pressuring NASA. Weiler was getting criticized from all sides. Once, while on a quiet cigarette break outside his NASA building, he encountered people on the street pestering him to restart the mission to Pluto.

By early November, the pressure caused Weiler to begin looking for a way out. In his search for a solution, he turned to Tom Krimigis at APL.

APL had not done many planetary missions—only one, in fact—but it had an impressive, decades-long track record of building and launching Earth observation and military satellites that performed well and were inexpensive. Moreover, their first and only interplanetary mission had been a resounding success. That mission had been born when Krimigis had helped spearhead NASA's development of the Discovery Program of small, competed, PI-led planetary missions. APL's Discovery mission was called NEAR, for "Near Earth Asteroid Rendezvous." The craft launched in 1996 and became the first spacecraft to orbit an asteroid, circling for a year around the asteroid Eros. Later, it would even land on Eros, a bonus achievement that was not even conceived of in the original mission proposal. And to top that, the APL team had completed the spacecraft *ahead* of schedule, and ended up doing the mission for $30 million *under* budget, giving the money back to NASA.

NEAR had been a triumph by every measure. How did APL manage to perform so well? One big part of it was a management philosophy of not having more managers than was absolutely necessary, since layers and layers of managers drove costs up. Instead, they put more responsibility at the level where the knowledge was—with their engineers. APL also kept their missions small through a fundamental desire to remain a lean organization. APL preferred to grow in prominence rather than head count.

All this had established APL, and its space department head, Tom Krimigis, as a team well known for being capable of delivering successful spacecraft missions on a tight budget. So Weiler asked Krimigis in mid-November if he could find a way to do the Pluto mission in a much less expensive way. Tom told him he could. "I can probably do this for a third of the cost of JPL. It's how APL works."

So with Weiler's encouragement, Krimigis got a small team to work intensely on the problem for a lightning-fast, ten-day prototype design and costing study. The team worked straight through the Thanksgiving holiday, and on November 29, 2000, Tom met with Weiler to give him the study results and cost estimate. Tom recalls:

> We essentially came up with what later became the New Horizons concept, including the shape of the spacecraft, paring it down to one plutonium battery—a leftover from Cassini—and other innovations to cut cost and to create a believable schedule to launch in time for a Jupiter gravity assist. Our study showed it, and I assured NASA that it could all be done for much less than $500 million, including reserves.

With APL's proof of concept, Weiler had found his path forward.

ABOUT-FACE

In late December of 2000, Alan got word that the ice was breaking and NASA was going to move forward after all on a Pluto mission. But how they would move forward was not at all how he expected. He had assumed that, if his campaigns of public and scientific pressure were successful, NASA would pick up the PKE mission where it had left off before the cancellation, select the instrument payloads as planned, and proceed to a new start. But on December 19 Alan got a call from a low-level insider he knew at NASA Headquarters: "You guys have won the battle; we're restarting on Pluto, but it's going to be your worst nightmare." His worst nightmare? What could that mean?

The next day Weiler publicly announced that, following the SSES's recommendation, NASA would try again to find a workable mission to Pluto, but this time the entire mission would be competed in one fell swoop: instruments, spacecraft, ground-operation plan, science investigations. Everything. It was to be done like the Discovery planetary exploration missions, the new mold of PI-led missions that Tom Krimigis had helped to give birth to, but this would be a competition for a much bigger prize—a much larger mission than any previous PI-led planetary project.

The Pluto competition would be open to all. So JPL, who had up to then thought they "owned" the Pluto mission, would now have to compete for it. A winning proposal would have to convincingly describe a mission that met three key criteria: it had to deliver on all the science that Lunine's Pluto Science Definition Team had said was a must (i.e., no skimping on objectives); it had to arrive at Pluto before 2020, even if forced to use its backup launch window; and it must do all this—from design to build to test to flight—on a breakthrough budget of $750M or less (in today's dollars, including credible budget reserves). This last was perhaps the tallest order of all—a budget that was barely half of what the PKE mission had been estimated to cost—and only about 20 percent of what Voyager had cost.

What made matters ultra-scary: proposals would be due on March 21; a schedule that seemed nearly impossible. NASA proposals for missions like this often run to a thousand pages of detailed design, comprehensive science, management plans, schedules, budgets, team bios, and more. Here proposers were being asked to cram what would normally be a year or more of work into just a few months to meet the March 21 deadline.

The day of Weiler's announcement, Alan's phone rang twice: APL and JPL were both putting together teams to compete for the mission. Alan was only forty-three years old, but after the gauntlet of the 1990s Pluto mission studies and politics, he was known as "Mr. Pluto," and someone who could effectively lead teams. Both major labs wanted him to head their proposals.

The call from Charles Elachi, by then the head of JPL, came less than an hour after Weiler's announcement. Alan spoke with Elachi, but held back on agreeing to lead a JPL proposal because he knew from Ralph McNutt that APL's Tom Krimigis would soon also be calling.

LEFT: Clyde Tombaugh, circa 1930, after discovering Pluto at age twenty-four. (Lowell Observatory)

BELOW: The 1930 telegram conveying the suggestion of Venetia Burney, an eleven-year-old girl from England, that the newly discovered planet be named Pluto. (Lowell Observatory)

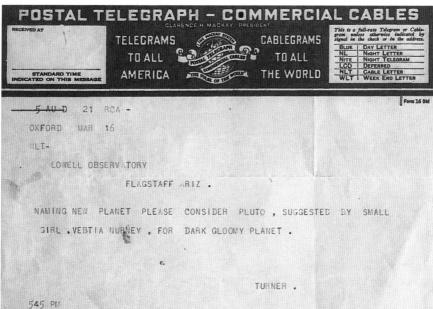

POSTAL TELEGRAPH - COMMERCIAL CABLES

TELEGRAMS TO ALL AMERICA

CABLEGRAMS TO ALL THE WORLD

5 AU D 21 RCA -

OXFORD MAR 16

WLT-

LOWELL OBSERVATORY

FLAGSTAFF ARIZ .

NAMING NEW PLANET PLEASE CONSIDER PLUTO , SUGGESTED BY SMALL

GIRL ,VEBTIA NUBNEY , FOR DARK GLOOMY PLANET .

TURNER .

545 PM

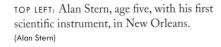

TOP LEFT: Alan Stern, age five, with his first scientific instrument, in New Orleans. (Alan Stern)

TOP RIGHT: Original Plutophile and New Horizons science team member Marc Buie. (© Michael Soluri/michaelsoluri.com)

ABOVE: Longtime New Horizons science team members Tiffany Finley, Leslie Young, Ann Harch, and Cathy Olkin (all front frow). (© Michael Soluri/michaelsoluri.com)

AT RIGHT: Original Plutophile and New Horizons science team member Bill McKinnon. (© Michael Soluri/michaelsoluri.com)

TOP LEFT: Science team member and deputy project scientist Cathy Olkin. (NASA/Bill Ingalls)

TOP RIGHT: New Horizons science team members Marc Buie and Hal Weaver. (© Michael Soluri/michaelsoluri.com)

LOWER LEFT: New Horizons Mission Operations Manager Alice Bowman. (© Michael Soluri/michaelsoluri.com)

LOWER RIGHT: New Horizons Project Manager Glen Fountain at the Pluto flyby. (NASA/Joel Kowsky)

ABOVE: Some of the New Horizons women engineers, scientists, and flight controllers photographed at Applied Physics Laboratory (APL) three days before Pluto flyby. (© Michael Soluri/SwRI/JHUAPL)

AT LEFT: Science team members Fran Bagenal and John Spencer at APL during the flyby. (Henry Throop)

BELOW: New Horizons science team member and geology team lead Jeff Moore discusses fresh Pluto data with team members during the flyby. (NASA/Bill Ingalls)

ABOVE: The NASA Outer Planets Science Working Group (OPSWG) in 1991. Alan and Fran (both seated) are in the center, Marc Buie is in the back row near center. (NASA)

BELOW: Rob Staehle (in front) and members of the Pluto Fast Flyby (PFF) design team, ca. 1992, at Jet Propulsion Laboratory (JPL) with a PFF mock-up. (NASA)

ABOVE RIGHT: New Horizons Mission Systems Engineer Chris Hersman with Alan Stern, taken near APL a few days before the flyby. (Alan Stern)

ABOVE LEFT: The "Pluto Not Yet Explored" stamp issued by the U.S. Postal Service in 1991, which was issued along with a set that showed pictures of the other planets. Rob Staehle and Stacy Weinstein took this stamp as a dare.

AT LEFT: The new U.S. Postal Service Pluto stamp set issued in 2016 after New Horizons had explored the planet.

BELOW: The external systems and scientific instruments aboard New Horizons. (NASA)

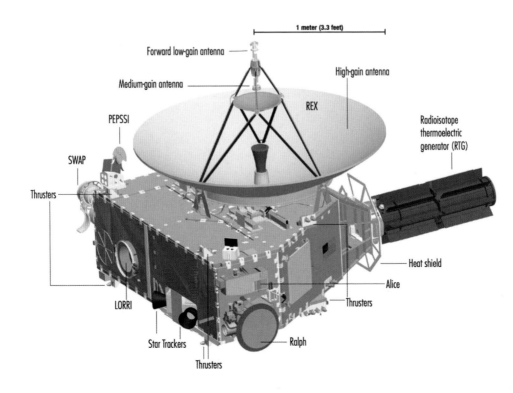

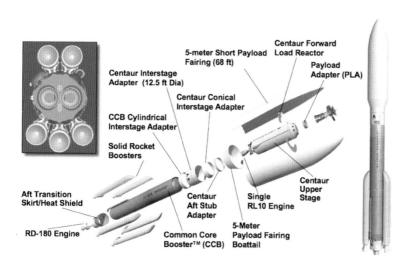

TOP: New Horizons during final assembly at its Florida launch site. (NASA)

BOTTOM: The augmented Atlas V rocket for New Horizons, showing all the stages and components described in the text, with a tiny New Horizons seen to scale at upper right. (NASA)

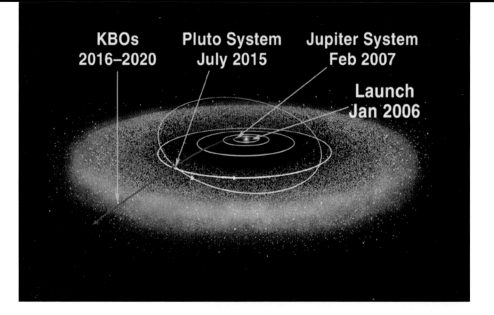

KBOs
2016–2020

Pluto System
July 2015

Jupiter System
Feb 2007

Launch
Jan 2006

OPPOSITE TOP: New Horizons mission leader Alan Stern in front of the spacecraft after its nuclear power generator was fueled on Friday, January 13, 2006. This is the last known picture of New Horizons; after this, the open hatch was closed for launch. (NASA)

OPPOSITE BOTTOM: Some of the New Horizons science team in front of their Atlas launcher about one week before it left our planet. (NASA)

THIS PAGE TOP: The route of flight of New Horizons across the solar system to Pluto. (JHUAPL)

RIGHT MIDDLE: The container, attached to New Horizons, carrying a portion of Clyde Tombaugh's ashes. (NASA)

AT RIGHT: Patsy Tombaugh (Clyde Tombaugh's widow) points to the sky shortly after seeing the launch of New Horizons on January 19, 2006.
(© Michael Soluri/michaelsoluri.com)

New Horizons leaps skyward atop the Atlas V rocket on a pillar of flame, January 19, 2006. (NASA)

TOP: Moments after the launch of New Horizons, David Grinspoon (foreground, right) celebrates with members of the science team and their families. (Henry Throop)

ABOVE: New Horizons flight controllers and others watching (and cheering) the launch from inside the Mission Operations Center at APL. (NASA)

A typical working shift at the New Horizons Mission Operations Center at APL during the nine-year flight to Pluto. (NASA)

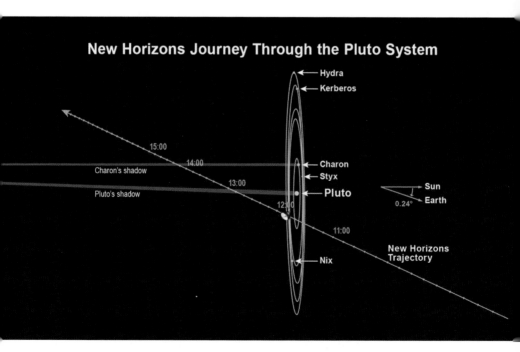

The route of flight of New Horizons as it sliced through the Pluto system on July 14, 2015. Times shown along the trajectory are Universal Time. (JHUAPL)

The special issue of *Science* magazine with results from the New Horizons flyby of Jupiter. (Reprinted with permission from AAAS)

12 October 2007

Science

Vol. 318 | No. 5848 | Pages 153–340

New Horizons at Jupiter

www.sciencemag.org

12 October 2007 | $10

Science

NEW HORIZONS
at Jupiter

AAAS

TOP: The flyby-day countdown to the moment of Pluto flyby at APL on July 14, 2015. The clock in the background shows minus 4 seconds. (NASA/Bill Ingalls)

BOTTOM: At the Pluto encounter, Alan, APL Director Ralph Semmel (center), and team members Leslie Young and Will Grundy (with Clyde Tombaugh's daughter, Annette, and son-in-law, Wilbur) hold up a just-cancelled version of the "Pluto Not Yet Explored" postage stamp. (NASA/Bill Ingalls)

LEFT: Engineers burning the launch-abort safety plan at the post–launch day party in Florida. (Morgaine McKibben)

BELOW: Burning the emergency communications plan at the team party the night of closest approach to Pluto. (Henry Throop)

TOP: New Horizons science team members react to their first look at the "fail safe" image of Pluto, early morning on July 14, 2015. (© Michael Soluri/michaelsoluri.com)

BOTTOM: Team members gathered around John Spencer's laptop at APL, gazing at the first hi-resolution image of Pluto. Back row: Jeff Moore, Randy Gladstone, Ron Cohen, Andy Chaikin, Bill McKinnon, Maria Stothoff. Front row: Laurie Cantillo, John Spencer, Alan Stern, Will Grundy, Steve Maran. (NASA/Bill Ingalls)

Alan also knew that the less experienced APL team would be the underdog in any competition for a mission of this scope. Yet he was wary of JPL's history of bloated Pluto missions, and he didn't really trust their commitment to see a Pluto mission through to completion on cost and schedule.

Anticipating both calls, Alan had put together a brief pair of questions to ask of both Elachi and Krimigis. First: "If I go with you, will I be your only Pluto PI?" This was crucial because Alan wanted to be able to scoop up all the A-team engineers and executives, with no competition for them to be on other Pluto proposals coming from the same institution. He wanted to make sure that any institution he went with had all their eggs in his basket, so to speak—that there was no way for them to win working outside his team's proposal. Alan's second question was "If we win, will you promise, in writing, that you will never let it go, that you will fight for it to the death if the mission ever runs into funding or political problems?" Alan:

> Both Elachi and Krimigis asked to sleep on my two questions and to call me back the next day. When Elachi called back, he spent half an hour explaining why JPL was going to beat APL by a mile, but JPL couldn't possibly deliver only one Pluto proposal, and why they couldn't promise to fight against NASA if it was ever canceled downstream. Basically, he spent the time on the phone telling me no to both my questions and trying to make me feel comfortable about going with JPL even though they would run multiple proposal teams with different PIs and would not promise to fight unconditionally for the mission if it later got into trouble. Krimigis called me back soon after and said, "Alan, you will be our only PI, and if we win it, we will never let it go. I'm giving you my word." I was very pleased with Tom's responses, but when I hung up that phone I thought, "I am screwed. JPL won't really back us, and although APL will, APL is the clear underdog in this and likely to lose to the stronger, more politically entrenched JPL." There was no easy decision.

One reason why APL was such an underdog was that, unlike JPL—who had historic success with two Pioneer flyby missions, two Voyager flyby missions, and both the Galileo and Cassini outer-planet orbiters—APL had absolutely no experience or track record of missions to the outer solar system. This mattered tremendously because there are numerous technical and management challenges unique to outer solar system missions. The travel times are much longer than those in the inner solar system, so spacecraft have to be built to last for years longer, and there are challenging operational logistics involved in operating spacecraft for those durations. The reliability and spacecraft "fault protection"—the ability of the spacecraft to automatically handle problems—have to be equal to the task of navigating in space for the years-long hauls between outer planets. The temperature extremes encountered require exquisite and reliable thermal engineering. Also, spacecraft operating at those greater distances from the Sun cannot power themselves with solar panels. They require nuclear power, which introduced another whole host of challenges, both technical and regulatory. Alan:

> I thought about the JPL-versus-APL decision long and hard that evening. I knew APL was up to the task, but it was a kind of Hobson's choice because going with them was riskier.
>
> I woke up in the middle of the night knowing the decision had to be to go with APL. I knew that I had to go with the team that really wanted it and would back it forever, but I knew that I would be making a choice that had real weaknesses. I also knew that if I went with APL, Elachi would consider me persona non grata at JPL for the rest of my career. It was sobering to choose APL, because if we did lose, which I thought could happen, the personal consequences of losing would be so high. But given what Elachi had said to my questions, versus what Krimigis had said, my choice just had to be APL.
>
> As I layed awake that night, thinking about the coming competition, I got more and more charged up by the challenge of

beating JPL. I went to work very early that morning to call both guys back with my decision. Krimigis was excited. Elachi was flabbergasted.

GOING TO WAR

Once Alan had agreed to lead APL's Pluto mission proposal, he and Tom Krimigis set about putting together a dream team. For the proposal and project manager APL brought in Tom Coughlin, their most experienced space project manager. Tom was also the man who had managed their successful NEAR planetary mission and brought it in $30 million under budget. In order to steer their Pluto mission through the perilous shoals of nuclear launch approval, they brought in APL's Glen Fountain, the coolheaded and brilliant head of engineering in APL's space department. Then Alan got to work choosing which scientists to invite to join the team as co-investigators.

Alan spent the Christmas holidays in 2000 at work recruiting a science team and working with APL on spacecraft design studies, handling a dozen top-level trajectory and spacecraft trades for how to architect the mission, and organizing team meetings to hammer out agreements on how the project work would be split up and what instruments would fly aboard the mission to explore Pluto.

What followed was a crazy period of seven-days-a-week, nearly round-the-clock work to design the mission in detail and to write up the phone-book-thick proposal addressing all of NASA's information requirements about it.

The pace was furious: decisions that would normally involve years of considered study were being made in days. It was thrilling, but every single decision was deeply consequential: any overreaching could produce a proposal that was unrealistic in its ability to meet cost and schedule or weight or power limitations; and underreaching could cost them dearly in how the review panels would rank the proposal against competitors. Alan felt he was riding a knife edge between two ways to lose, and he knew that he had to create a finely balanced proposal that was technically and managerially perfect, because any flaw could be seized on to justify a win by the more experience competitor teams at JPL.

The work to build, review, and perfect an entire mission proposal by NASA's deadline in March continued week after week, and weekend after weekend. But what happened next was simply Kafkaesque: In early February 2001, just as the assembled draft proposal had passed its first complete review, the brand-new Bush administration released its first federal budget. Shockingly, and despite NASA's just-announced competition for Pluto missions, the budget zeroed all NASA funding for a Pluto mission and instead made the Europa mission a new start! Within a day or two of that, NASA aborted the Pluto proposal competition.

Alan was incredulous. And furious. And he suspected the hand of JPL, who stood to gain from the competition's demise:

> I was so mad I couldn't see straight, and I smelled something fishy. If Europa went forward, JPL would be guaranteed to get the work, because that mission had simply been assigned to JPL—without competition—and it was also a far bigger monetary prize than winning Pluto would be.

Alan speculated that JPL had worked behind the scenes to persuade the Bush administration more or less to trade the Pluto mission for a new start on Europa. He also believed that JPL had another interest in killing Pluto, because if APL actually won, APL's hand would forever be strengthened as a powerful competitor in all future outer solar system exploration.

Alan immediately called Krimigis. He recalls Tom saying something like, "It's time to break some legs." He had never heard a space scientist talk that way. "My God," he thought. "I definitely chose the right guy. He's going to war for this mission!"

Tom decided to fight fire with napalm, calling his political ace in the hole—the powerful senator Barbara Mikulski of Maryland, where APL is located, and then chair of the Senate funding committee responsible for space exploration. At Tom's behest, Mikulski wrote a sharp letter to NASA, demanding that NASA resume the Pluto mission competition. Her letter scolded NASA, reminding them that by cancelling the competition they were stripping the power of the U.S. Congress to decide on whether to fund a Pluto mission. She told them they could not do this. As the chair of the Senate committee that held NASA's purse strings, NASA had no choice but to listen. NASA resumed the competition.

The game was back on.

"WHATEVER IT TAKES"

Aside from Alan's team with APL, four other teams were preparing Pluto mission proposals to NASA. Two of the most formidable teams were from JPL, with more senior, and more-experienced principal investigators, each famous veterans of Voyager and other legendary missions. One team was led by Larry Soderblom, a widely respected planetary geologist from the United States Geological Survey, who had been point man for studies of icy satellites on the Voyager camera team and who was a darling of JPL upper management. The other was led by Larry Esposito, a planetary polymath, principal investigator of the ultraviolet spectrograph aboard NASA's Cassini Saturn orbiter, and a professor of planetary science at the University of Colorado (which also meant that Alan was going up against one of his former grad-school professors). Alan knew he was too inexperienced to match the track records of these giants of the field; his team would simply have to turn in the best proposal. Alan often thought of the competition as if he and his team were David, going up against multiple Goliaths.

Enter budding planetary scientist Leslie Young, who had been a part of the discovery team of Pluto's atmosphere when she was an MIT undergraduate back in 1988. By 2001 though, Leslie had a Ph.D. under her belt and had come to work for Alan as a postdoc.

Leslie became a key part of the proposal-writing team. Brilliant, and brimming with enthusiasm, she put in epic amounts of work and even led a key piece of the proposal on her own: to make the proposal credible, NASA required each team to show a flyby plan that actually fit in all the needed observations within the capabilities of the proposed spacecraft design. And it wasn't enough just to show that the planned scientific instruments had the right resolution, the right sensitivity, and all the other technical specs to do the observations being promised to meet (or exceed) the mission objectives. The team also had to demonstrate that the spacecraft design and instrument capabilities proposed, along with their chosen flyby trajectory, could conduct all the needed observations in a flight plan that fit together, showing that it could work seamlessly, that there was enough time between turns, not too much power drawn at any given time, never exceeding the available amount of data storage, and so on in many other respects.

Creating this flyby plan was like a chess game in ten or more dimensions. Leslie led that development, and in doing so she became a world expert in this kind of complex mission planning. At first, Alan was a little concerned that a young postdoc like Leslie could lead what he knew would be such a complex effort. But Alan saw all the right qualities in her. And when he asked her to sign up for nights and weekends of work for the rest of the proposal effort, Leslie told Alan, "I'm here to win. It's whatever it takes." That phrase, "whatever it takes," became both Leslie's mantra and made such an impression on Alan that it became a rallying cry for the project whenever the going got tough.

More than one hundred people became involved in the APL proposal effort. A few of the key people included Alice Bowman, an experienced APL flight director who was put in charge of designing how New Horizons would be operated across its ten-year journey; Chris Hersman, an electrical and system engineer who was put in charge of overall New Horizons design; and Bill Gibson, of Southwest Research Institute (SwRI), their most experienced space project manager, who was tapped to corral together the design, building, and testing of all seven scientific instruments across four separate corporations and universities, on cost and on budget.

In addition to tackling all the engineering and management design challenges and trying to write a spotless proposal, Alan wanted to put the proposal through an unusually intense gauntlet of internal "red-team" reviews to find and remove all of the technical, managerial, and even pedagogical problems in their proposal. Most proposals of that era planned an intensive red-team review like this, where a mock review panel of experts would assess the proposal critically and find weaknesses, but Alan wanted three red teams to rise to this higher standard as an antidote to their more experienced competition at JPL. That would be expensive and time-consuming—and it was borderline manic. APL resisted due to workload and cost issues, but Alan prevailed. Alan:

> For a while I wasn't very popular within the proposal team because I demanded so much—so many reviews, so many revisions, so many long nights and long weekends. I wasn't in it just to propose. It was win it or go home, make or break, with no prize for second place.

HOUSTON, WE HAVE A NAME

One seemingly minor but vital task while writing the proposal was coming up with a name for the proposal and the mission. As PI, Alan was responsible for picking the name, but he wanted buy-in from the larger team.

Anyone who has ever started a band is familiar with this process. You want to come up with the perfect name, but so many get rejected that after a while they all run together, or they all start to sound equally bad.

Alan wanted a name that was both descriptive and inspiring. This being NASA, of course, lots of acronyms were suggested. Given that it was a Pluto mission, they all had *P*'s in them, and there were lots of *E*'s for "exploration" or *M*'s for "mission." Names came and went: dull, forgotten acronyms like COPE, ELOPE, POPE, and PFM. Some slightly better ones came in, such as PEAK ("Pluto Exploration

And Kuiper-Belt"), or APEX (for "Advanced Pluto EXploration"). But none was particularly inspiring, strong, sufficiently catchy, or memorable.

Along the way, Alan's team learned that one of their competitors—the proposal being written by Larry Esposito at the University of Colorado, in concert with JPL—was to be called POSSE for "Pluto Outer Solar System Explorer." This was a fine descriptive name, but Alan thought it lacked inspiration. He joked, "Who are they looking to arrest?" Alan wanted something more hopeful.

After dozens of tries with acronyms, Alan realized he was going to have to break out of the NASA acronym mold. He decided that instead of a conjured acronym, he wanted a good name that was in itself a short inspirational phrase or slogan, one that captured the essence of what they would achieve with the mission.

Again there were many suggestions. Someone suggested it simply be called "X" in honor of Tombaugh's original search for "Planet X," and for the futuristic feel of that name that hinted at NASA's pioneering X-planes, like the X-15. Other suggestions that came in were "New Frontiers" or "One Giant Leap." But something was wrong with each of these: to some, "X" connoted the drug Ecstasy, and "New Frontiers" referred to Kennedy's space program, something Alan feared the Bush administration, then in office, would chafe at. As to "One Giant Leap," in honor of Apollo's "one giant leap for mankind," he was afraid their proposal would be ridiculed as "one giant leap of faith." Time was slipping by: with every passing week, Alan was barraged with the plea, "We need a name. We're already red-teaming the proposal and we have no name for it. Get us a name!"

Then, one rare Saturday back home in Boulder, Alan was taking a run and working out ideas in his head. His thoughts turned to the naming dilemma. Alan:

> I decided then and there that the very positive word "new" should be part of the name, because what we were doing was new in so many ways. Damn, I thought, "New Frontiers" was soooo close, but it's got political baggage. Then, as I waited for a streetlight to change so I could run on, I happened to look to the

Rocky Mountains on the western horizon, and it just hit me. We could call it "New Horizons"—for we were seeking new horizons to explore at Pluto and Charon and the Kuiper Belt, and we were pioneering new horizons in how to run the first-ever PI-led outer-planets mission. Nobody, I thought, could find a black cloud connotation lurking in a bright name like New Horizons. New Horizons was easy to say, easy to remember, and it symbolized that the mission would be doing something new in two important ways. I could tell it was right as I ran farther. I tried to shoot it down in my head, but I couldn't find a way. By the end of the run I was settled on it, and I recall thinking that it was in a way a historic decision: "*If* we win this competition, and *if* Congress finds the funding to make this mission happen, and *if* the mission actually reaches the launch pad, and *if* all these things go our way and we successfully explore Pluto—then the name New Horizons will be found in textbooks and encyclopedias for centuries."

INVENTING NEW HORIZONS

To win, a number of features of the New Horizons proposal were intended to differentiate it from its more experienced competition. The core of that was an expanded instrument payload that promised to achieve all of NASA's required scientific observations at Pluto, but which was supplemented with other instruments adding new dimensions to the science. Alan felt that even though some of this was not required, it would broaden the mission and bring in whole new communities whose support he might later need to keep the mission funded. Adding these bonus instruments was possible largely because APL had a track record of spacecraft and mission development that was much less expensive than what the JPL mission studies had said was possible, in turn opening a budget wedge for the additional scientific gear.

The proposed New Horizons payload was centered around the integrated imaging and spectroscopy package that Alan's team had put together for the (now-canceled) PKE competition.

First was PERSI (Pluto Exploration Remote Sensing Investigation): a powerful set of cameras and composition spectrometers in the visible, infrared, and ultraviolet parts of the spectrum. PERSI would photograph the surfaces of Pluto and Charon in detail, seeing details small enough to pick out features the size of city blocks. It would also make infrared observations that would map what Pluto and Charon's surface materials were made of. In the ultraviolet part of the spectrum it would reveal the structure and composition of Pluto's atmosphere and search for an atmosphere around Charon.

Next REX (Radio-Science Experiment) would probe the pressure and temperature of Pluto's atmosphere as a function of altitude, a requirement by NASA. To build REX, Alan landed Stanford professor Len Tyler's team—the most experienced radio-science team that had worked on Pluto mission radio-science-experiment development in the past and that possessed the most capable technology in the world for this type of experiment. By bringing in Tyler's Stanford team, not only was Alan locking in a highly capable radio-science group, he was gaining a major strategic advantage over all competitors.

Next he selected two instruments for charged particle observations (Ralph McNutt and Fran Bagenal's area), called PEPSSI (Pluto Energetic Particle Spectrometer Science Investigation) and SWAP (Solar Wind Around Pluto), designed to study the composition and escape rate of gases escaping Pluto's atmosphere.

The cherry on top of the New Horizons payload was LORRI. LORRI stood for LOng Range Reconnaissance Imager. It was a simple black-and-white camera, but it was fed by a big, long-focal-length telescope that would add three very important aspects to New Horizons science, none of which any JPL mission had ever really planned for. First, with LORRI's high-resolution telescope, New Horizons could obtain images with five or more times the resolution that NASA required for the mission's Pluto and Charon maps. As a result, LORRI would vastly amp up the game for geology results and provide the kind of detailed images that first planetary flybys never had before, with resolutions good enough to detect features the size of buildings rather than the size of city blocks. Second, because it was a high-magnification imager, LORRI would also allow New Horizons to beat the Hubble Space Telescope's best Pluto resolution for ten weeks on approach and ten weeks on departure—meaning they could turn a fast flyby "weekend at Pluto" into a multi-monthlong science visit—a huge bonus that would allow for many new kinds of research to be done. Third, and perhaps best of all, LORRI's high-magnification imaging made it possible for a single spacecraft flyby to obtain basic maps of the far sides of Pluto and Charon, the sides that would not be visible except from afar during a single flyby.

This instrument suite packed a lot more punch than what was needed to meet NASA's minimums, but proposing it also required a degree of shrewd salesmanship. The New Horizons team had to simultaneously present the payload as accomplishing much more than the minimums, but also as entirely feasible to build on budget and on schedule—lest the proposal be deemed overreaching.

To accomplish this, the New Horizons proposal was careful to demonstrate the simplicity of each instrument and the risk-lowering "heritage" of each. That is, how each was built on designs from previous space instruments with which the team had direct experience, whose performance had already been previously proven in space.

At the same time, the proposal pointed out that there were many instruments that could have further enhanced the science but that the team chose not to include, such as a magnetometer, which would search for Pluto's magnetic field. This part of the win strategy—arguing they had actually denied themselves many other good ideas—had the feel of a high-stakes poker game in which they were trying to read and outsmart the competition by offering to do more with less.

In addition to this spectacular instrument set, New Horizons also proposed several mission innovations to make their proposal harder to beat.

First, the proposal offered to achieve a very fast trip to Pluto. How? In addition to planning the Jupiter gravity assist, which would shave almost four years off the flight time, they also added a simple but reliable solid rocket stage to the giant Atlas V launcher, which would cut the travel time to Pluto. Overall, the proposal offered to make the journey in just eight years, arriving in mid-2012, if they could launch by the December 2004 Jupiter-Gravity-Assist (JGA) launch window, or nine years if they had to launch in 2006—the backup launch window that was the last chance for a JGA for many years. The New Horizons team argued that this faster trajectory also lowered risk, because the shorter mission timeline meant less time for something to go wrong. The faster flight, they pointed out, also meant an earlier arrival, reducing the additional risk of an atmospheric freeze-out happening before the craft got to Pluto.

The team also proposed several ways to make the spacecraft extremely effective at Pluto. The flyby was going to be fast, there was no way around that. The spacecraft would whip past Pluto at more than thirty thousand miles per hour, and all the most crucial observations would be made in a matter of just hours. So the design included the capability to run up to five instruments simultaneously during the flyby, and vastly increased the onboard solid-state flash memory to be able to store a whopping thirty-two times the data harvest that PKE had promised to gather. The design also made it possible for the spacecraft to execute very quick turns back and forth between Charon and Pluto, packing even more science into the choreography of the many observations planned for the day of closest approach.

In order to afford all these capabilities, the proposal had to be very clever about making offsetting cost savings in other parts of the spacecraft and the mission to keep it all within NASA's budget box. The most ingenious of these savings was the plan to have the spacecraft go into "hibernation" during a large part of the trip, meaning that the spacecraft would shut down most of its systems for years between Jupiter and Pluto, with only minimal communication and navigation capabilities remaining active. Such hibernation had not been done by any NASA mission before (though some Pluto mission studies had also included it), but importantly, it promised to greatly reduce mission-control staffing costs because only a skeleton crew would be required to maintain contact with the spacecraft during hibernation. As a result of this and other mission-operations innovations, the team planned to run the flight mission with a staff of fewer than fifty people—a dramatic breakthrough compared to the 450-plus that it took to run Voyager. Also, New Horizons chose to purposely reduce the telecommunications capability of the spacecraft to ten-times-lower bit rates at Pluto than Voyager had at Neptune. This allowed a smaller, lighter-weight, lower-cost antenna, and a lower-power transmitter, letting the craft get by with just one nuclear power supply rather than two—all saving even more power, cost, and mass. The idea was that as long as they could maximize data collection at Pluto, and reliably store the data on board the spacecraft, then they could take their time returning it all to Earth. The proposal team adopted the mantra "If you can take almost a decade to fly it to Pluto, then surely you can take one more year to get all the data back."

The New Horizons team also found several important ways to minimize the risks of a low-cost mission to the very edge of our planetary system. Sending only one—rather than two spacecraft as had all the previous first missions to planets—was an inherently risky venture, but they could not afford two. So the proposal countered by making all the active spacecraft systems "fully redundant." From propulsion to onboard memory storage to flight control and guidance computers, to twinned power systems and twinned telecommunications transmitters and receivers—every crucial component had a fully functional backup that could be used in the event of a malfunction.

Some past Pluto mission studies had sacrificed such redundancy in the name of weight or cost savings. The New Horizons team made it a selling point, in effect telling NASA that if the space agency was going to fly the mission, New Horizons was going to make sure it was capable of getting the goods even if there were some hardware failures along the way. The New Horizons proposal's scientific payload also included redundancy. With eight separate imaging cameras, two different spectrometers that could record composition, two separate plasma instruments, and two REX radio experiments, the design lowered risk by being able to cover all key science goals even if any given instrument failed before or during the one and only Pluto flyby.

PLAYOFFS

The NASA competition to win the Pluto mission was set up in two stages. Anyone who could amass a team and the necessary financial resources to propose could enter the first round of competition. NASA would then "down-select" only the two best proposals from the first round, and those two teams would enter a more elaborate and detailed head-to-head final round of competition. In a way being a finalist is like the difference between the regular season and the playoffs in pro sports because only one team goes home with the trophy.

But then, unlike in sports, the losing Pluto proposal team wouldn't be playing again next year; in fact, they would never get another shot.

Five teams registered with NASA to compete in the first round. Of those, one underdog team dropped out mid-stream, but on April 6, 2001, four teams crossed the finish line, turning in massively detailed technical and management proposals outlining how they would build and fly a Pluto mission. NASA then convened large panels of experts in every technical subspecialty—in project management, in budget analysis, in risk analysis, and a dozen other areas, to rank and evaluate the four proposals. Their in-depth review process took two months.

When the announcement of NASA's down-select decision came, Alan was in Paris, at an international Kuiper Belt meeting. Many in the Pluto/Kuiper Belt community were there and knew that the announcement was imminent. Near midnight on the evening of June 6, Alan got back to his hotel near the Arc de Triomphe. As he was walking through the lobby a clerk at the front desk said, "Mr. Stern, you have four phone messages." (This was, of course, before the modern smartphone era.) Alan looked them over but none of them were from NASA Headquarters. One, however, was from a Ms. "Yung," with a Colorado area code. Alan recognized the number and realized it was actually from Leslie Young, but he thought it was about a research paper they were just then finishing. He almost let it go until morning, but then decided to return her call. Leslie picked up the phone, but all Alan could hear was background pandemonium. Leslie shouted over the partying crowd in Boulder to Alan, "We've been selected to be one of the two finalist teams!"

New Horizons made it! Well, at least they were in the finals, but the next round of competition would be even tougher.

The other winning effort was the POSSE proposal from JPL, with principal investigator Larry Esposito from the University of Colorado (CU). So the Pluto mission competition had come down to Boulder versus Boulder, CU versus SwRI, in a crazy three-month final race to develop detailed plans for the mission.

Esposito's team was formidable, and their mission proposal was excellent. The POSSE team also had the weight and track record of JPL on their side, with the highly experienced Lockheed Martin corporation as their spacecraft builder. The New Horizons team still felt like it was the David in this battle, but at least now it was only up against a single Goliath—one team or the other would prevail.

The POSSE proposal differed from New Horizons in many respects. Because JPL and Lockheed were institutionally more expensive than APL, POSSE needed to use a lower-performance launch vehicle with no third stage to fit in the cost box. As a result, their flight times to Pluto were longer. Additionally, their spacecraft was heavier and had not made conscious trades like a lower-performance telecomm system to save cost. And POSSE also took what Alan considered "sucker bait" and proposed to develop new technologies, like tiny high-performance thrusters that, while scoring points for technology development, cost money and introduced the risk that these new technologies might not be ready on time, or that their cost might escalate. Additionally, POSSE had loaded their spacecraft up with eleven scientific instruments, falling across the bad side of what Alan considered the "Christmas tree line"—with too much promised for a breakthrough small budget and short development timescale.

With the news that New Horizons had been selected as a finalist, so began three more months of incessant travel between Colorado and Maryland, new, more-detailed design analyses, cost analyses, instrument performance analyses, backup planning for a possible 2006 launch, red-team reviews, and late nights working seven days a week. The team members' home away from home became a Sheraton hotel in Columbia, Maryland, barely ten minutes north of APL. They spent so many late hours working in the bar there, over beers, with laptops open, scribbling on napkins, that the hotel staff knew them all by name. One of the bartenders, Linda Lappa, became a regular at their after-hours meetings and was affectionately added into some of the unofficial project-management charts, shown as "project bartender."

The final few weeks before the September 18 deadline for "best and final" proposals were an insane sprint. As Leslie Young often said, there would be time for sleep after the proposal was turned in.

On September 10, 2001, the team conduced their final review of the materials that would go to print for signatures before being shipped. But the next morning, only a week before proposals were due, came the terrible shock and tragedy of the terrorist attacks of September 11. Everyone who is old enough to remember that day will recall where they were and what they were doing when they heard the news about terrorist attacks in New York and Washington. Anxiety choked the air, especially so in Maryland, only a short distance from Washington, DC, where one of the planes had hit. All air travel was shut down. APL—itself a defense lab—was completely evacuated due to bomb threats.

Like everyone else in the country, the New Horizons team found themselves in 9/11 shock, but the NASA deadline was only a week away. So the team evacuated from APL and moved the final work on the proposal to a marathon session at the Sheraton where they rented conference room working space on the spot. Alan recalls, "It was hard to go on during such a tragedy, but we had to keep going. I think that somehow our national pride, combined with the fact that we were working on trying to create something historic, in the face of so much meaningless destruction, helped to motivate us that terrible week."

Ultimately, with all business in the nation virtually shut down over the tragedy, NASA gave both teams an extra week to complete their work, extending the proposal deadline to September 25, and both teams made it.

With the finalist proposals in hand, NASA began an even more rigorous set of technical, cost, and management reviews to ferret out the strengths and weaknesses of the two competing teams and their approaches to exploring Pluto.

In the final stage of every mission competition like this, NASA does a "site visit" with each team, during which NASA's panel of experts grills the proposal team in an intense, daylong oral exam that involves a detailed presentation of the team's proposal. Every aspect of the proposal is examined: the team, the institutions involved, the design, the budgets, the management team, the detailed schedules with thousands of events to reach the launch pad, and, fundamentally, the science that will be accomplished. For New Horizons, the oral exam came on October 16. The New Horizons team spent the prior two weeks reviewing and critiquing one another's presentations, doing dress rehearsals, even bringing in a panel of outside experts to simulate the grilling NASA would do.

Alan felt pretty good going into their orals. He knew his team was ready and that their proposal was as flawless as any he'd ever been involved in. And personally, after more than a decade of work on various versions of this mission and the intense marathon work on the New Horizons proposal, he felt ready for pretty much any questions that might be thrown at him during the site visit. But he was also reflective of the stakes. Alan:

> I remember thinking as I drove to the airport to fly to APL for the site visit, "This could be the last Pluto mission trip I ever take. I've been doing this for twelve years—since that first step into NASA Headquarters to see Geoff Briggs back in May of 1989. And it all comes down to this."

The entire New Horizons proposal team—almost one hundred engineers, scientists, managers, and others—assembled in a large auditorium at APL for the NASA site visit, along with a dozen or so corporate executives from APL and SwRI, and NASA's review panel of twenty or so experts.

The daylong visit was grueling. Then, after all of the technical and management presentations, and NASA's detailed questioning, there was a tour of APL's space department, where the review panel could see the design, test, and mission-operations facilities that would be used to support New Horizons.

Then, to close, the presenters and panel reassembled in the auditorium, where Alan was to give a five-minute soliloquy, the final impression the review panel would be left with. Alan reiterated why Pluto had to be explored and why New Horizons had the right team and was the right mission for NASA. As the room darkened in a final dramatic flair, Alan put up a last slide—showing New Horizons flying by a detailed artist's impression of Pluto created by scientist-artist Dan Durda. Then, just as Alan was finishing, asking the review panel to recommend New Horizons to go forward, something unexpected happened.

"Just as I finished that closing pitch," Alan recalls, "and the lights came up, I thought I saw the review panel chair wink at me. From where he was sitting I knew that no one else could see him do that. I just about fell over. I thought, 'Did he just indicate "good job" or did he just indicate he thinks we'll win? Did that really happen?'"

WHEN GOLIATH FELL

Late that November, the planetary science community was gathered for their biggest annual meeting—"the DPS"—a planetary nerdfest, mentioned earlier, as a pivotal annual gathering for science, socializing, and politicking about the planets and their exploration. That year, Alan was in charge, and the meeting was held in New Orleans, his boyhood hometown.

On Thursday, the twenty-ninth, just as Alan was leaving a technical session for a coffee break, Tom Morgan, a NASA Headquarters executive, came up to him and said, "See that pay phone? There's a phone call there for you." Alan:

We had heard that NASA was to announce the Pluto mission winner that week. So I knew Tom was saying, without saying it, "This is your call—you're about to find out if you won or lost." I walked up to the phone and said a little prayer to myself, because I knew I was seconds away from getting NASA's verdict—and one with no appeal.

It was Denis Bogan on the line from NASA Headquarters—NASA's Pluto program scientist. I said hello, and he matter-of-factly said, "Alan, we've concluded our evaluations."

Time slowed down. I thought: my career's most important endeavor was to be decided at a pay phone in the middle of a noisy coffee-break crowd buzzing with conversations. It came down to this one moment; whatever Denis would say to finish his sentence would be IT. And then he said:

"Congratulations. We've selected New Horizons to be our Pluto mission."

I felt a shiver run up my spine! We had beat JPL—Goliath had been vanquished. When I got off the phone I excitedly ran to a computer to write a message to the entire team. It read simply: "We did it! NASA Headquarters just called to say we won the Pluto competition and will be funded to proceed! More soon." Then I ran out into the crowd of a thousand or more scientists to find Tom Krimigis and whispered to him the news. Tom grabbed me in a hug and we literally started dancing together, right there in the halls of the conference. No one knew what the hell we were doing or why, and we got some pretty funny looks.

That night, the members of the New Horizons team that were in New Orleans for DPS walked in a throng down Bourbon Street, past open bar doors with music spilling out onto the street. Alan thought a lot that night about growing up in New Orleans, of being a kid in the 1960s dreaming of becoming involved in space exploration. Now it was Stanley Kubrick and Arthur C. Clarke's iconic, futuristic year of 2001, and he had won the chance to explore the farthest worlds humans had ever attempted, and he was home in the town where all his ambitions to be a part of space exploration had begun. It was an amazing confluence that he savored all evening and into the next day.

That night the New Horizons team and a large group of well-wishers ended up at a big dark bar off Bourbon Street with a three-piece band playing in the corner. For the next several hours, that gaggle let their hair down and, frankly, just got completely shit-faced listening to the band—drunk with joy, laughter, relief, and excitement for the long adventure to come.

5

NEW HORIZONS AT LAST?

"YOU WON, BUT YOU LOST"

Alan left New Orleans the next day and headed home to Boulder. The following week a letter arrived at his office from NASA's then science chief Edward Weiler, making the win official. Alan shut the door, opened the envelope, and began reading:

> Dear Dr. Stern:
> I am pleased to inform you that your concept study report, "New Horizons: Shedding Light on Frontier Worlds," submitted in response to the awarding of a Phase A study contract for the Pluto-Kuiper Belt (PKB) mission, has been selected to proceed.

If only the letter had stopped there. But as Alan kept reading, his smile straightened and then was lost altogether. The next paragraph began:

> However, there are a number of requirements that must be met before NASA will continue proceeding. . . .

The letter went on to describe what were, in effect, a gauntlet of ways New Horizons could be canceled by NASA. To avoid that, first, it had to stay on track to launch in order to meet the last 2000s Jupiter launch window. Second, the budget cap for the project's total cost couldn't be raised, so any overrun would prove disqualifying. The letter also listed numerous categories of milestones, which had to be met on schedule and a requirement to successfully navigate the complex maze of nuclear launch approval. Weiler's letter didn't offer any help, or even encouragement, it just described a minefield of obstacles—and made it clear that any single one could result in cancellation.

Alan had won many other NASA projects by then, and never had he received any award letter like this. The letter's tone indicated that NASA didn't believe they could make the schedule, or get the project funded by the Administration, or get nuclear launch approval in time. "I read that letter three times, and I sat down and I thought, 'My God.'"

Also buried in Weiler's letter was a launch delay from December 2004 to January 2006—from the next to last to the very last Jupiter-gravity-assist launch window of the decade. This delay was a mixed blessing. It gave more time for all the work and planning and building and approvals necessary to be ready for launch, but it also meant there would be no backup window: if the team missed 2006, there would be no other gravity assist for a decade. And that delay also had a more insidious implication on the mission budget. Those additional 13 months that extended the project to 2006 would add expenses that in turn would require stretching the budget even thinner because they would have to carry the engineering team longer, making it all the more likely to go over budget—crossing one of the Weiler letter's boundaries for cancellation.

When Alan later shared the letter with a colleague not on the team, the assessment was just as dispiriting: "You won, but you lost. You'll spend the next year or two or three working on this, but most likely fail on one of Ed's stipulations, and then you'll be canceled. It might have been better to have lost, as Esposito and Soderblom did; at least they won't have to sink the next few years into this."

BUSHWHACKED

A few months later, as the project was just spooling up, in early February 2002, Alan found himself on one of the many public outreach trips he took for New Horizons, this time in New Mexico, at a school-wide assembly of the Clyde Tombaugh Elementary School to give a talk about the budding project. After he finished talking to the students and answering their questions, his host, planetary scientist Reta Beebe, took him aside.

"Have you seen President Bush's new budget that came out today?"

"No, why?" said Alan.

"New Horizons has been canceled."

Alan was in disbelief. That couldn't be right. It was just awarded by NASA, a part of Bush's very own executive branch! Alan:

> Reta and I went straight to her office and got online, and I found the NASA language in the president's proposed budget, which had been released that morning. It was true—the Administration had zeroed our budget in the next fiscal year. Amazingly, it stated that the Pluto mission was cancelled "due to cost overruns."

Alan's jaw hit the floor. They weren't over budget. Indeed, how could there be cost overruns when the project didn't even have a formal contract yet? Was this just double-talk revenge for forcing the competition to completion via Senator Mikulski the previous year? Or was it some misunderstanding inside the Bush White House that Pluto Kuiper Express, cancelled back in 2000 had been in overrun and now it was canceled? But if that was so, how did New Horizons become the victim? Where did this move even come from? Perhaps the guys from the Office of Management and Budget (OMB), who seemed obsessed with funding Europa exploration and had been gunning for Pluto, had found another way of killing off Pluto exploration. Or had JPL, knowing now they had lost the proposal competition, worked to see that if they couldn't win it, then the entire effort would be scuttled? None of it was clear except that Reta had been right: New

Horizons had been canceled, and their job to stay on schedule now included a tough budget battle ahead.

Thankfully, Senator Barbara Mikulski, as the chair of the Senate's Appropriation Committee for NASA, stepped in again, providing bridge funding for the next year. But for future years, the Senate language made clear, funding would be dependent on what the next planetary exploration "Decadal Survey" decided about the priority of exploring Pluto over other missions like Europa.

That Decadal Survey was, as its name implies, a new, once-in-a-decade review of all of NASA's planetary mission priorities, undertaken by the National Academy of Sciences. It is a hugely influential shoot-out in which representatives of all the different areas of planetary science, and advocates for different kinds of planetary missions, make their cases and then agree on a consensus ranking of priorities for which missions are to be funded and launched for the entire decade to come.

Following Senator Mikulski's quick intervention, Alan heard from Ed Weiler. The Bush Administration had agreed that they would support New Horizons if, and only if, it was ranked at the top of the Decadal Survey. And Weiler made it clear, saying in effect: "You can't just be on the A-list; you can't even be number two on the A-list. You have to be ranked as the number one mission priority. If you are not, then it will be over. Period."

That was a tough challenge—that New Horizons had to get past the Decadal with more than just a recommendation for funding, they had to be #1 on the runway for priority. And then this news came: the National Academy declared that, due to conflict of interest, all key members of the New Horizons team were prohibited from serving on the Decadal Survey panels. The rationale: they were conflicted by having their funding at stake. Perhaps so . . . But this meant the most knowledgeable and passionate advocates for Pluto exploration were barred from having direct input into the very process that would decide the fate of the Pluto mission.

It was all a huge challenge. Why? Because, first, there was an intense competition in the Decadal Survey; a lot of missions were vying to get funding priority. Second, unlike New Horizons, many of the other missions being considered had not yet been selected, so they had the advantage of making unreasonable promises—basically "Christmas-treeing" their benefits to increase their appeal. Third, the advocates for those missions—like a Europa orbiter or future Mars rovers—were able to serve on the Decadal Survey's panels because their missions hadn't been selected and people hadn't been named to the teams: they hadn't been funded and so weren't seen as conflicted in the way the New Horizons advocates were. It was maddening: "I felt like we were trying to drive a car without our hands on the wheel," Alan said.

CLYDESDALES

During the same period that the Decadal Survey was being conducted, the intense work of designing the New Horizons spacecraft was moving forward aggressively, involving a growing army of engineers and scientists. New Horizons was also going through a mountain of work to prepare for NASA's Mission Confirmation Review (MCR) and doing it in record speed and on a record low budget. But the Bush administration's cancellation and the review by the Decadal Survey was taking a huge amount of time and energy and had cast a pall of uncertainty over the future of the mission.

Ordinarily, when a NASA mission is heading toward its Confirmation Review, the project team can count on a great deal of help from NASA Headquarters, with many kinds of technical and logistical support. But, given the at best tentative funding state of the project, the New Horizons team was more or less stuck with going it alone. Another problem in getting support was then NASA Administrator Sean O'Keefe. O'Keefe had come to NASA from the Office of Management and Budget where he had been when OMB had tried to kill off the Pluto mission in favor of one to Europa. After arriving at NASA, O'Keefe made it clear that he was no friend of Pluto or New Horizons.

Senator Mikulski's bridge funding gave New Horizons enough oxygen to run for a year. But during that time the players in NASA were not enthusiastic about New Horizons, because it wasn't a Bush Administration–supported mission. The result for New Horizons was more than frustrating—it was just bizarre. Alan:

> It was a weird "Alice in Wonderland" time. My project manager, Tom Coughlin, who'd done so many space projects no one could remember all their names, one day called me and said, "Alan, I have never been on a project like this. Normally, all the horses at Headquarters are pulling for you. They're on your team, but not on this one. The Clydesdales aren't even hitched to our wagon, in fact it seems like the Clydesdales are just staring at us thinking, 'What's the wagon even for?'"

DECADAL DECISION

Yet again, the exploration of Pluto and the New Horizons team was in a situation of "win big or go home." And all the while the project had to play two fronts at once: while the team worked tirelessly to design New Horizons in 2002 and early 2003, they also had to wage an intensive campaign to make the case why the exploration of Pluto and the Kuiper Belt should be the very highest priority of the Decadal Survey. Project members spoke to individual members of the Decadal Survey, wrote scientific white papers for its committees, worked on getting positive articles in the press, and encouraged the public to weigh in; they even once again enlisted the Plutophiles of The Planetary Society to help them build support.

That June, on the night before the press conference at which NASA and the National Academy of Sciences were going to announce the results of the Decadal Survey, Alan got a call from a journalist who was close to people in Sean O'Keefe's office. Apparently the journalist had received some leaked information and told Alan, "You're going to get what you want tomorrow, but not quite the way you expected."

Recalling something similar that he'd been told in December of 2000 when the Pluto mission was being resurrected, Alan had thought:

"What the hell does that mean?" and then he chuckled to himself, "Why is this beginning to sound familiar?" Alan:

> The next morning, I was in my office early because Ed Weiler wanted to speak at 7:30 A.M., and I knew it must be about the decision of the Decadal Survey. I was at my desk when the phone rang. It's sort of like that phone call I got in New Orleans the day that we won the proposal competition: you know in the next sixty seconds something big is going to be settled one way or the other. Weiler said hello, and then he said, "The Decadal Survey has ranked the exploration of Pluto as its top priority, and the Administration is going to go along with that."
>
> "Wow," I thought to myself. "We finally are going to have clear sailing after all these years." But then I remembered the journalist's caution of the night before, and sure enough, no sooner had Weiler shared the good news about the Decadal's decision than he continued: "But there's one other thing."
>
> What Ed said was that NASA wanted to add another rocket stage to New Horizons, a high-tech ion-propulsion stage using solar energy to add even more speed and shorten the trip time. And they wanted to have JPL build it. "Don't worry about what it costs," Weiler said, "we'll cover it." But I thought, "What is this? We don't need this stage, or any of its complications."

It was so laden with unnecessary complication that Alan believed that it was a JPL or JPL-Weiler ploy to hamper JPL's then arch-rival APL. And it didn't make any sense. Why? First, with a fast launch like New Horizons, the spacecraft wouldn't be near the Sun for very long, so solar energy simply couldn't power the ion stage effectively for more than a matter of perhaps a year. Second, although Weiler was saying its cost would not come out of the New Horizons budget, the funds still had to come from Weiler's own limited budget, and it would be expensive. Alan and Glen ballparked the number at $300 million, maybe more. Alan:

The third reason was that it suddenly put JPL back in the driver's seat. These were the guys who were at the helm of every past, stillborn Pluto mission, who lost the competition to us, and then seemed to try to get us killed because, we suspected, "if they can't have it then no one can." Apparently, now, JPL had found a new way back in. If that stage cost too much, or if it weighed too much, or if it wasn't ready for the vehicle launch approval in time for the Jupiter launch window—anything—we would never get launched.

The whole concept had risk written all over it for New Horizons, so Alan came up with the only plan he could think of to save the project from this new potentially fatal detour. Alan:

On the phone with Weiler I said, "I hear you, Ed; we're very glad to be at the top of the Decadal, and we'll study how to add the electric propulsion for you." Then I put the phone down and made a plan to completely circumvent adding that electric propulsion stage boat anchor to New Horizons by running out the clock on Ed, knowing that eventually NASA would have to drop the solar-electric-ion stage in order to make the only launch window of the top-ranked Decadal mission.

Alan got the team together and, a couple of days later, made a call to NASA Headquarters. He said New Horizons wanted to start the solar-electric study, and then provided NASA with an impossibly long list of data items that they needed in order to begin. Alan:

We more or less created a ridiculous homework assignment for them, probably months of work, but we only gave them a month to do it. When the deadline came a month later, of course the information we asked for was not complete, and we used every incomplete item as a reason to send them back with even more homework: define this better, finish that in more detail, et cetera. . . . We basically kept them in the mode of never being ready to design that electric stage we didn't need. And we knew we would win by running out the clock because the Mission Confirmation Review was coming in the spring, and with the Decadal now firmly supporting us, NASA could not stop New Horizons over the lack of readiness of this unneeded add-on propulsion stage.

And that is exactly what happened. In the spring of 2003, New Horizons had its Mission Confirmation Review. The team had, over the past two years, successfully passed each precursor technical review, scoring A's across the board at their nonadvocate cost review, their preliminary design reviews, and their systems requirement reviews—all of them. But the MCR was "do or die"—the gate every mission must pass to begin building their spacecraft.

The New Horizons' MCR review was held at NASA Headquarters in Washington in March of 2003, and New Horizons passed: *without* the ion stage.

Nearly fourteen long years after it all began in 1989, a mission to Pluto was now approved for construction and its funding was finally secure. Behind them now, at long last, was the seemingly endless era of studies, funding battles, and politics. Ahead was a spaceflight project— building, launching, and flying New Horizons to explore Pluto and the Kuiper Belt—the final planetary frontier of our solar system.

ALABAMA

After the Decadal Survey battle was won and after killing off Weiler's cumbersome solar electric-propulsion stage, New Horizons seemed finally out of its endless political battles. But there was still one final twist ahead: now that the mission was past MCR and approved to be built, JPL raised the question with NASA of which NASA development center would manage New Horizons. And because all of NASA's planetary missions back then were managed by JPL, JPL volunteered to be responsible to also handle New Horizons.

When Tom Krimigis and Alan got wind of that, they saw that JPL was in effect offering NASA to put themselves in charge of the very mission that had beat them in the Pluto mission competition. Beyond the obvious conflict of interest, Alan and Tom suspected that JPL management saw New Horizons as an existential threat to JPL's more or less monopoly on outer solar system exploration for NASA. Losing the NEAR asteroid mission to APL had been a blow to JPL back in the 1990s, but NEAR was a pretty simple, close-to-home planetary mission and JPL learned to stomach competition in the "minor leagues" of close to Earth planetary exploration. If JPL now lost their franchise on the outer solar system—the far away "big leagues" of planetary exploration, and had to sit on the sidelines while APL built NASA's mission to the farthest world in history, then JPL would soon likely have to compete for every future planetary mission.

Alan and Tom tried to talk NASA Headquarters out of putting their former rival in charge of the project office, explaining the conflict of interest, but to no avail. So once more they had to turn to their champion, Maryland's powerful Senator Mikulski, who negotiated with NASA Headquarters to have the NASA New Horizons project office set up on neutral ground, at NASA's Marshall Space Flight Center in Huntsville, Alabama. Once again, Mikulski had saved the day. Alan:

> Someday there really should be a "Mikulski Crater" or "Mount Mikulski" named on Pluto or Charon. She sure earned it.

6

BUILDING THE BIRD

The full New Horizons team that designed, built, and flew the mission involved more than twenty-five hundred men and women. Alan often referred to them as the "Corps of Exploration," a name inspired by Lewis and Clark's intrepid company two centuries before New Horizons.

About half the total New Horizons force worked on its launch vehicle: a two-stage Atlas V rocket and its custom third stage. About a third more of that workforce designed and built the spacecraft and the science instruments, and planned or executed the mission operations. The remainder were employed on nuclear launch approval, on the science team, on public outreach, and other efforts.

This workforce stretched far beyond SwRI and APL, with more than a hundred participating companies and universities, plus NASA, and other government agencies in the mix. Major subcontracts under APL, SwRI, or NASA included Ball Aerospace, which built the "Ralph" camera spectrometer instrument; JPL, which provided the Deep Space Network that would keep New Horizons in contact with Earth; Lockheed Martin, which provided the giant Atlas V rocket;

Boeing, for the third-stage rocket that the mission needed to supplement the Atlas V and speed New Horizons toward its Jupiter encounter; Aerojet (now Aerojet Rocketdyne), for both the spacecraft propulsion system and Atlas V solid rocket boosters; and Honeywell, which produced the gyros that would help New Horizons stay oriented in space.

Organizationally, the project was led from Alan's office at SwRI in Boulder, with his staff organized under an "Office of the PI." But most day-to-day engineering and mission operations fell to APL, which designed and built the spacecraft and operated its mission control. SwRI led the development of the seven scientific instruments and developed and staffed the mission's science operations center.

APL's project boss during the proposal and early design/build stages was Tom Coughlin; when Tom retired due to health problems in late 2003, Tom Krimigis asked Alan for his pick to replace Tom Coughlin. Alan asked for Glen Fountain.

Glen was a highly experienced project manager and veteran of many space missions at APL, and he'd forged a close and trusting relationship with Alan. Like Clyde Tombaugh, Glen grew up in a small Kansas farm town and wound up exploring the farthest reaches of the solar system.

During the proposal phase of New Horizons, Glen was running the engineering branch of APL's space department, and he helped orchestrate the mission's technical development. "Basically," Glen recalls, "Alan came and lived down the hall from me for three months while we wrote that proposal." Later, after New Horizons was selected, Glen became the point man for "nuclear launch approval"—navigating New Horizons through the labyrinth of regulatory hurdles required to launch a spacecraft powered by plutonium.

At SwRI, Bill Gibson was both payload and project manager, responsible for corralling all seven scientific instruments through design, development, and testing and also responsible for day-to-day budgets, schedules, and subcontract management. Gibson was SwRI's most experienced spacecraft project manager and a gifted people person, whose quiet Southern accent helped calm even the most stressful decision meetings.

At both SwRI and APL, New Horizons, the sexy "first mission to the last planet," had been able to have the pick of the litter for both the engineering and mission operations team leads, recruiting an all-star team of talented and dedicated managers in the prime of their careers, with deep technical experience and unrelenting drive. The backbreaking, four-year design/development/test/launch schedule demanded nothing less. It also demanded intensive travel, work on nights, weekends, and holidays, and all-out commitment to perfection, since there was no tolerance in the schedule for "redo's" and no second spacecraft or launcher if New Horizons failed.

MAINTAINING THE LINK

As we described before, New Horizons had the intense challenge of pulling off its mission on a lot less money than its predecessor, Voyager—five times less on an inflation-adjusted basis—and this required some tough thinking on how and where it made sense to save money, and where *not* to.

Inventing a new spacecraft costs a lot of money because every part and process must be tested and proven to ensure that it will survive the rigors of the long trip on its own. Both to save money and to increase reliability, New Horizons borrowed electronics designs from previous APL planetary missions, avoiding starting from scratch whenever possible. For example, APL more or less cloned the spacecraft command-and-data-handling system from its just completed MESSENGER and CONTOUR spacecraft, and SwRI built the "Alice" ultraviolet spectrometer largely following an Alice ultraviolet spectrometer design it had used for the Rosetta comet-orbiter mission.

One area where New Horizons made important advances was the telecommunications system, the vital radio link between Earth and the spacecraft, which manages the transfer of information in both directions—commands from Earth to the spacecraft, and data and health reports ("telemetry") from the spacecraft back to Earth. As we described earlier, the New Horizons team made a decision to reduce the telecommunications capability at Pluto because antennas are heavy and deep-space transmitters are power hungry, and also because reducing the communications capacity saved money to help fit in NASA's budget. But that meant that it would then take more than a year after the Pluto flyby to get all the precious data on the ground. Alan told the team: "If we don't fit it in the cost box, we're not flying. I know you'd rather have faster bit rates, but if you want to actually get to Pluto, we have to meet NASA's cost target, and that means making compromises."

The clever design decisions used to create the frugal, lightweight telecom system for New Horizons illustrate just one example of the many decisions made, from propulsion to guidance to data storage to thermal control, to create an outer planets spacecraft that broke the mold on cost.

CHOOSING A ROCKET

To get to Pluto as fast as possible, the New Horizons team had to build a very light spacecraft and buy a powerful rocket. That combination (along with a Jupiter gravity assist) was the recipe for the highest possible speed to cross the solar system. At the time New Horizons was started, the United States didn't have a powerful enough rocket that was already operational, but two powerful new rockets were in development. Lockheed Martin was building a massive new launcher called the Atlas V, and Boeing was building another new rocket called the Delta IV. Both were to be enormous: over two hundred feet tall and capable of generating millions of pounds of thrust at launch.

Boeing and Lockheed Martin were fierce competitors, fighting to win contracts for every launch. In 2002 and 2003 when New Horizons was being designed, NASA and the New Horizons team went through a careful process to choose between the two rockets, comparing, for example, how each required different spacecraft mounting, and each had different performance specs, acceleration and vibration environments, and costs. In the end, Atlas V won the day over Delta IV, in part because the Atlas was likely to be ready earlier and had more launches planned to prove itself before 2006 arrived.

The most powerful version of the Atlas V, called the 551, was selected to power New Horizons on its way. The first stage of this monster rocket is 107 feet high and 12.5 feet in diameter and is centered around a powerful, Russian-built engine that burns liquid oxygen and kerosene. Attached to that first stage are five gigantic solid rocket motors, which fire in tandem with the first stage. Together they would loft New Horizons (and the rocket's upper stages) to hypersonic speeds of over ten thousand miles per hour. Atop that was a second stage, called Centaur. It's forty-two feet high and is powered by an American-made Aerojet Rocketdyne RL-10 engine that provides 22,300 pounds of thrust. Significantly, the Centaur stage can be started and stopped multiple times, which was necessary to put New Horizons onto the right course to Jupiter.

Encasing both the Centaur—and New Horizons atop it—is the rocket's "fairing," or nose cone, designed to protect the spacecraft from the fierce wind forces generated by launch. New Horizons ordered the lightest-weight Atlas 551 nose cone, which saved weight and further improved the 551's launch performance.

Yet even this maxed-out Atlas V 551 still wasn't powerful enough to loft New Horizons on its way to Pluto. Centaur could put New Horizons into Earth orbit, and then take it out into an orbit that would carry it as far as the asteroid belt beyond Mars. But to achieve the performance to reach Jupiter and then Pluto, it was necessary to add a custom-built third stage atop the Atlas V. For this, the New Horizons team selected a highly reliable, well-proven solid rocket fuel third stage, called a STAR 48. That stage, which would fire for a brief eighty-two seconds, would accelerate New Horizons to almost 14 G's and make it the fastest spacecraft ever launched, capable of reaching the orbit of the Moon ten times faster than the Apollo missions did and flying on with that speed for almost a decade to cross the 3 billion miles to Pluto.

TAKING PLUTONIUM TO PLUTO

How does one power a spacecraft that will be traveling for at least a decade on a journey so far from the Sun that our star shines there at less than a thousandth of its brightness at Earth? Solar arrays won't work that far from the Sun, and no battery is powerful and light enough to do the job of powering a decade-long mission. But the radioactive decay of plutonium (an element that was discovered in 1940 and was named for Pluto) passively generates heat without fail—and that heat can be turned into electricity. For this reason, plutonium-fueled nuclear batteries have been the power supplies of choice for deep-space interplanetary missions to the most distant planets from the Sun. But carrying a plutonium-fueled nuclear battery introduced its own complexities: some technical, but others political and regulatory.

NASA, along with the Department of Energy and the Department of Defense, perfected, tested, and flew many plutonium power batteries for just this purpose beginning in the 1960s. These devices are called Radioisotope Thermoelectric Generators, or RTGs. RTGs are cylindrical in shape and about the size of an oil drum; because the plutonium generates so much heat, RTGs are equipped with cooling fins. RTGs have two primary jobs: one is to power their spacecraft, and the other is to contain their plutonium in the event of a launch accident.

The plutonium used in RTGs is packaged in small pellets made of plutonium dioxide. These pellets are in turn clad in iridium and bottled up inside the RTG's black graphite casing.

As an RTG creates heat from the radioactive decay of its plutonium, that heat gets turned into useful power through simple, completely passive devices called thermocouples. Thermocouples have two sides, one inside the RTG that's hot, and one on the outside of the RTG, facing space, that's cold. The temperature difference across the thermocouples generates an electrical current, which is what powers the spacecraft. The amount of heat generated by the RTG on New Horizons is about five kilowatts, and from that it produced about 250 watts of electricity to power the spacecraft when it launched.

RTGs are extremely reliable. They produce power at a steady but slowly declining rate and can be operated for decades. Their decline in power with time is due to the radioactive half-life of the plutonium. For New Horizons, the 250 watts its RTG generated at launch declined to about 200 watts a decade later at the time the spacecraft reached Pluto.

In 2001, when NASA announced it would conduct a competition for missions to explore Pluto, the Agency owned two spare RTGs from the development of its Galileo mission to Jupiter and its Cassini mission to Saturn. NASA offered to provide the winning Pluto mission provider with either of these RTGs.

Once selected, the spare RTG for New Horizons had to be disassembled and fully inspected (after all, it had been in storage for over ten years) and then rebuilt. Lockheed Martin served as the contractor to refurbish the RTG; the Department of Energy's Los Alamos National Laboratory then prepared the plutonium to fuel it.

While that work was going on, there was an elaborate, parallel effort that all RTG missions undergo to assess the risks of a launch accident, including the preparation of a comprehensive environmental impact statement, which addressed those risks and showed that they had been reduced to acceptable levels. This rigorous approval process involved forty-two state and federal agencies, up through the State Department and a sign-off by the White House.

Glen Fountain supervised this complex process. Glen:

> We managed it knowing that we didn't have the seven or eight years it would normally take. We had four, and that was a tough challenge.

Using an RTG solved the problem of how to power a mission going so far from the Sun, but it also posed its engineering challenges for the designers of New Horizons.

For example, the RTG is heavy, weighing more than 125 pounds. So during launch the structure holding it has to support its weight multiplied by the G-forces the rocket stages for New Horizons produced—up to 14 G's at maximum acceleration, making it weigh fourteen times what it normally would on the ground. So the joint holding the RTG had to be able to withstand a force of 14 times the RTG's normal weight. This challenge was made even tougher by the heat the RTG creates, which acts to weaken the metal in the support joint. So New Horizons engineers had to design the joint to be strong enough to support its weight against the G-forces during launch, even when the joint was hot.

A second engineering challenge RTGs pose to spacecraft designers is that the radiation the plutonium produces is bad for spacecraft electronics. As a result, all of the systems aboard the spacecraft had to be designed and tested to withstand this radiation, as had been done for previous RTG missions, like Voyager, Galileo, and Cassini. That in turn added complexities and cost to the development of the spacecraft, but there was no choice: without an RTG there would be no way to power New Horizons when it was far from the Sun.

EYES, EARS, AND EVEN A SENSE OF SMELL TO STUDY PLUTO

As the principal investigator, Alan was in charge of all aspects of the New Horizons project. All the teams of scientists and engineers, as well as others such as public relations and project management, ultimately reported to him.

One key support role requiring both expertise and diplomacy was the project scientist position. Every NASA science mission has such a position, and larger missions like New Horizons also have deputy project scientists. These scientists are responsible to the PI for representing and negotiating the interests of the PI and the larger science team in the seemingly endless day-to-day stream of meetings that are needed to coordinate a complex space mission.

The New Horizons project scientist position was filled by a highly talented, affable, and diplomatic planetary scientist named Hal Weaver.

He was selected because he is extremely broad scientifically, and conversant in all the different sublanguages of all the subspecialties that make up the New Horizons science—from geology to surface chemistry, and from atmospheric science to plasma physics. Hal was also selected in part because he was a seasoned experimentalist who understood the engineering and operations of the many kinds of scientific instrumentation used on New Horizons.

Hal had not been part of the original Pluto Underground. His background was instead primarily in the study of comets. But Hal had long been fascinated with the Kuiper Belt as the region of the solar system where certain comets originate. He had known Alan since the 1980s and they had collaborated together on some research. Hal describes his role as project scientist this way:

> The role is to be the PI's right-hand person on the project, to keep the PI informed about what's happening on the ground, and as design or test or other issues get raised, to be the scientific voice there to help the engineers figure it out. My job was to be the scientist who knows the engineering too, so that when the engineers tell me, "This is really hard," or "This is going to cost a lot to implement," I can help them find a design that not only meets the scientific objectives but also meets the project's mass, power, cost, and schedule needs.

When humans send scientific sensors to distant worlds, we're launching proxies for our eyes and our other senses. This is most obviously true with cameras, which allow us to "see" landscapes where no human eye has actually been. But it is also true with other kinds of instruments, which allow us to "listen" to the vibrations of distant magnetic fields and "sniff out" the gases in an alien atmosphere, to discover what these landscapes are made of, what lies beneath their surfaces, and what hidden forces, flows, and fields can tell us about the history and nature of other worlds.

A key design decision that was made early on New Horizons was that the spacecraft would not include a "scan platform," a movable turntable that points the cameras and other instruments in different directions without having to move the entire spacecraft. Although platforms increase the flexibility of the observations that can be done during a flyby (e.g., allowing the cameras to point at the planet while the antenna is pointed in a different direction back to Earth), they add weight, complexity, and cost. Voyager and other high-budget outer planet missions used scan platforms, but at one-fifth the cost of Voyager, New Horizons simply could not afford this luxury. Not having a scan platform in turn meant that all the instruments aboard New Horizons would be "body mounted" on the spacecraft, so pointing instruments to view their targets would mean pointing the entire craft for each observation.

Everyone involved in designing New Horizons knew that, unlike the first flybys of Venus and Mars and Jupiter, there were no plans to follow up its exploration of Pluto with orbiters or landers. The data they would collect would need to suffice for the foreseeable future as the complete body of knowledge about Pluto and its moons.

But because New Horizons was built in the 2000s, technology had greatly advanced and it was able to bring advanced capabilities that the Mariner and Voyager teams of the twentieth century had not been able to include: new sensors, much faster data-collection capabilities, and much greater instrument sensitivity. All seven instruments New Horizons carried were more advanced than anything that had been brought to bear on a previous first planetary flyby. Next we'll describe the instruments that New Horizons carried to study the Pluto system.

We'll begin with the "Alice" ultraviolet spectrometer. Picture a spectrum of the wavelengths of light that our eyes can see, going from red to blue. But what's bluer than blue? Ultraviolet. Those wavelengths, beyond what humans can see, reveal the composition of atmospheric gases. The details of the Alice instrument give some sense of how far instrument technology advanced from the days of Voyager to New Horizons. The Voyagers had also carried ultraviolet spectrometers, which had two pixels, or picture elements, so they could observe two separate ultraviolet wavelengths at once. This meant that building up a useful spectral map was a slow and time-consuming process of sweeping those two pixels across all the necessary wavelengths needed to build up a spectrum, and then laboriously pointing the instrument boresight at location after location, repeating this process to map the spectrum across the disk of each flyby object. In contrast to Voyager's old-style ultraviolet spectrometer, Alice contained thirty-two thousand pixels, so it could observe at 1,024 wavelengths at each of thirty-two adjacent locations—simultaneously, vastly speeding the process of obtaining ultraviolet data over what Voyager could accomplish.

Closely married with Alice, is the "Ralph" instrument. Its name came from a silly joke referring to the old *Honeymooners'* TV show characters, Ralph and Alice Kramden. Whereas Alice's objective was primarily to study Pluto's atmosphere, Ralph's objective was to map and also determine the composition of Pluto's surface. The size of a hat box, Ralph contains two black-and-white cameras, four color filter cameras, and an "infrared mapping spectrometer" to map surface compositions. Ralph can see colors that are redder than any red humans can see, at wavelengths which are called infrared, where minerals and ices have characteristic spectral features that can be used to reveal the surface materials at any given location in Ralph's field of view. Ralph's spectrometer splits infrared light up into 512 spectral channels from 1.25 to 2.5 microns. Again, a comparison to the historic standard for flyby exploration, Voyager, is illustrative. The Voyagers' equivalent instrument, called "IRIS," was about the same size as Ralph, but because it was designed and built with 1970s technology, it contained just one infrared pixel. By comparison, Ralph's mapping spectrometer has sixty-four thousand pixels. So on the Voyagers, IRIS's telescope had to be pointed at each place on a target body to get a spectrum of just that place, and then repointed to successive locations to slowly build up a spectral map. But Ralph obtains a spectrum at each of 64,000 locations all at the same time—blanketing a target to map all its locations simultaneously—something light-years ahead of what Voyager could do.

Determining the temperature and pressure of Pluto's atmosphere was another objective of New Horizons. To make these measurements,

New Horizons carried REX, which stands for Radio EXperiment. REX was designed to function in essentially the opposite way from its more primitive counterpart on the Voyagers. The Voyager radio experiment worked by sending radio X-band (4-cm wavelength) waves through planetary atmospheres it flew by from the spacecraft toward Earth. These waves were picked up by groundbased antennae of NASA's Deep Space Network. By measuring the ways in which that radio signal was altered by passing through the atmosphere of the various planets and moons on Voyager's itinerary, scientists could determine temperatures and pressures in those atmospheres. Because Pluto's atmospheric pressure was much lower, this technique wouldn't work, so REX solved this problem by performing the experiment the other way around: the Deep Space Network would blast a far more powerful signal than any instrument aboard a spacecraft could—with tens of kilowatts of radio power. Then REX would receive and record the signals sent from Earth that had passed through Pluto's atmosphere.

In order to measure atmospheric temperature and pressure, REX compares the frequency of the radio waves passing through the atmosphere to those of a reference standard. The shift that it measures is proportional to the bending of those radio waves caused by their passage through the atmosphere it is studying, which can in turn be used to compute the atmosphere's pressure and temperature. In addition to determining atmospheric pressures and temperatures, REX is also capable of measuring the temperature of surfaces it stares at.

Capping off the "remote sensing" instruments on New Horizons— the sensors that observe Pluto and its moons through optical and radio telescopes—was LORRI, which stands for the LOng Range Reconnaissance Imager. LORRI is essentially a magnifying telescope feeding a megapixel camera. Unlike Ralph's color and spectroscopic capabilities, LORRI takes only black-and-white images. But because its telescope is of a much higher magnification than the one on Ralph, LORRI's images have much higher resolution and show much greater detail. LORRI's higher resolution also allowed New Horizons to see features on Pluto and its moons from much farther away than with Ralph. As a result, beginning about ten weeks before the flyby, LORRI would be capable of seeing more details on Pluto than the Hubble Space Telescope ever could. With this capability, LORRI also allowed New Horizons to map *all* of Pluto—even the parts they didn't fly directly over on flyby day. Remember: Pluto rotates slowly, taking 6.4 Earth days to turn around once on its axis. This means that as New Horizons was approaching Pluto, the last time it was only able to see the "far side"—the side not flown over at closest approach—is 3.2 days before the flyby. New Horizons would be millions of miles away when it got its last look at that other hemisphere, but with LORRI's telescope, New Horizons would be able to get good imaging resolution on those terrains.

The next two instruments that New Horizons carries are so-called plasma instruments. Plasma is the term planetary scientists use for electrically charged particles. This is the domain of planetary science that experts like Fran Bagenal and Ralph McNutt study, and it is the hardest to describe for non-experts because it involves things people don't normally encounter in our daily lives. On Pluto, plasma is created as sunlight ionizes the gases in Pluto's atmosphere. As a result, by studying that plasma it's possible to determine the rate that Pluto's atmosphere is escaping and what the escaping gases are made of.

The two New Horizons instruments designed to study plasma are SWAP (Solar Wind Around Pluto) and PEPSSI (Pluto Energetic Particle Spectrometer Science Investigation). PEPSSI measures very high-energy (megavolt) charged particles. PEPSSI can reveal the composition of material leaking out of Pluto's atmosphere. Is it carbon? Is it oxygen? Is it nitrogen? Is it something else? SWAP's job is to measure the escape rate of Pluto's atmosphere, and it does that in an interesting way. As Pluto's atmosphere escapes into space, there is a place in front of Pluto at which the outgoing, escaping gas reaches a pressure balance with the Sun's incoming solar wind, and there is sort of a standoff where the two flows are balanced. The faster an atmosphere is escaping, the farther out into space it reaches that point of pressure balance with the incoming solar wind. So finding how many thousands of kilometers from Pluto that occurs allows one to determine the escape rate of gas from Pluto's atmosphere. That's SWAP's job.

There is one final instrument on New Horizons. It is called the SDC, or Student Dust Counter. SDC counts the impacts of interplanetary dust particles (tiny meteoroids) onto its detector surfaces. Every time there's an impact on the face of the dust counter, SDC creates a little voltage jump in the instrument that reveals how massive the impacting particle is. SDC's job is to measure interplanetary dust at much greater distances from the Sun than any other dust impact detector sent into space. The farthest any dust detector had operated from the Sun before New Horizons was just shy of the distance of the orbit of Uranus, or about half of the way to Pluto. SDC would provide a continuous trace of the solar system's dust density all the way out from Earth to well beyond Pluto.

SDC was also the first instrument built by students to be carried on any planetary mission. Getting it approved was not easy but Alan felt strongly from the start of the project that students should have opportunities to participate, so he used the idea of an educational training opportunity to convince NASA to add SDC to New Horizons. Today, thanks to that pioneering role of New Horizons in carrying a student-built instrument, most NASA planetary missions carry a student instrument, and they are now considered valuable tools for training the next generation of planetary explorers.

ALLIGATORS IN THE WATER

Throughout 2002 and 2003, the New Horizons team raced to design, build, assemble, and test the spacecraft and its payload, to pass numerous NASA technical and cost reviews, to work toward nuclear launch approval, to build a mission control, and to undertake all the other processes that had to dovetail in order to be ready to launch by the critical Jupiter launch window in early 2006. To complete the spacecraft required building literally hundreds of separate components for the guidance system, the communications system, the propulsion system, the ground systems, all seven instruments, and more. Every aspect of the spacecraft had to pass its own tests and then be shown to work perfectly together.

By early 2004, perhaps inevitably, there came a time of crisis as some components failed testing, and others fell behind schedule. This isn't unusual for spacecraft projects, but New Horizons really didn't have much wiggle room to make up for these problems with schedule adjustments because of its one-shot January 2006 launch window.

About that same time, NASA's New Horizons program office at the Marshall Space Flight Center in Huntsville, Alabama, assigned oversight of the project to a brash and talented young manager named Todd May. Todd, an engineer by training, came from the human-spaceflight world and had little prior experience in robotic spaceflight. Alan and Glen were skeptical that Todd would be able to be much help, given his lack of experience in the many skill sets required for robotic planetary exploration.

Immediately on taking the job, Todd told Alan that he wanted to come and visit SwRI Boulder, SwRI San Antonio, APL in Maryland, and other sites where work was being done on New Horizons so that he could meet all the principals and get the lay of the land for each facet of the project's status. Alan:

> I remember our first conversation. Todd said to me in his thick Alabama accent, "I want to come out and get to know you better, so I'm coming to Boulder to talk. Let's pick some dates." I didn't know who Todd May really was, and I thought he was just going to be some NASA manager who pushes a lot of paper and files a lot of reports. I kind of figured, "We're in trouble on half a dozen aspects of the project, and I really don't have the time to babysit this guy, but he's the boss NASA assigned; he's already been to APL (where I'd met him briefly) so I'll have to make time for his Boulder visit." When Todd came to visit me, he also visited Ball Aerospace, which is also located in Boulder, where they were building the Ralph instrument. Then Todd made four or five more trips in rapid succession, visiting APL, then other key project participants, getting himself immersed in New Horizons. And it didn't take him very long, maybe a month,

before he realized just how tough some of our developing problem areas were.

For his part, Todd May didn't know anything about Alan Stern
either, or New Horizons, when he was given the role of managing the
mission for NASA. Todd:

> One of my early trips on New Horizons was to APL. There was
> a monthly review going on, and that's where I met Alan and
> Glen. I sat through the review, and then Alan gave a little sci-
> ence overview. When he did that he did what he always does:
> he started with a PowerPoint of the stamp of Pluto that says NOT
> YET EXPLORED. I've got to tell you, that hit me where it counts.
> That's the kind of exploration that really gets me excited. The
> whole idea of learning some truth or making some discovery or
> going somewhere that we haven't been—to me that's just a big
> part of who I am. You could say he had me at "not yet explored."
>
> That day, I saw a project risk chart that the APL folks put to-
> gether. They had about five or six risks that were all pretty high
> likelihood and that could cause the mission to fail. I told the
> team, "You guys don't look like you're on a trajectory to be suc-
> cessful."
>
> About a week later, I got a call from Mike Griffin, who at that
> time had just taken over from Tom Krimigis as head of the APL
> space department. And Mike started by telling me, "You're say-
> ing to people that we aren't on a path to success. Are you trying
> to get this mission canceled? Are you trying to get me fired?" I
> said, "No, sir. I'm trying to make you successful, but I'm telling
> you, you've got a bunch of development risks, and I'm worried
> about that."

Todd was truly concerned, and called for a deeper review of the
project—a look at every aspect of every spacecraft subsystem, every
one of the seven instruments, the flight software, the ground system
(mission control), the launch vehicle, the RTG, and the nuclear

launch approval. He wanted to find out for himself where all the project areas of concern were, independent of what the SwRI and APL teams had identified. So he formed a team of experts in each area, who literally spent hundreds of hours on deep-dive cost, schedule, and technical reviews. In what Alan called a ninety-day-long "proctology exam," Todd determined that most things were on track, but that there were deep problem areas as well.

The biggest problem Todd's reviewers found was with the Ralph instrument, which was very much the centerpiece of the scientific instrument payload. Ralph was responsible for accomplishing more of the scientific objectives at Pluto than any other instrument onboard. Ball Aerospace was having major difficulties developing it. As a result, the instrument was behind schedule and far over budget. In aerospace, classified projects have higher priority over civilian ones, and Ball kept changing out key personnel on Ralph, taking the better engineers over to other projects. Every effort Alan, Glen, and Bill Gibson made to get Ball to quit poaching engineers, to get back on schedule, and to rein in cost increases failed. For example, as the cost escalated, the APL/SwRI team looked for ways to simplify Ralph's design. But Ball said that there would be a steep bill for the simplification since it meant redesign and reanalysis of many aspects of the instrument, and that it would actually end up costing New Horizons more. Alan once told Todd about it: "I feel like we're in a hostage situation with Ball. We can't launch without Ralph, and they know we'll have to pay whatever it costs, no matter how high it goes."

Todd's review found other problems, too. Nuclear launch approval was being held up at a number of agencies. Plutonium fuel production was also behind because of a work shutdown at the Los Alamos National Labs (more on that in the next chapter). Costs were also increasing for the spacecraft's propulsion system, and the development of the special-purpose third stage to top out the Atlas V rocket was running late.

Alan remembers at the end of Todd's review process a very painful conversation about how many "alligators in the water" the New Horizons team was fighting. Alan:

Todd basically told us, "My team has looked the project over a lot deeper now, and I'm pretty convinced you guys really aren't going to make it. You are submerging the various problems by trying to solve them yourselves. You need more money and some serious come-to-Jesus persuasion of some of your contractors by us at NASA, and you haven't been asking for either." Then Todd told me, "I'm going to tell NASA Headquarters that you can't make it as is."

For anyone who had been involved in NASA projects, "alligators in the water" was a clear reference to the gators that lurk in the ditches, streams, and marshes around Kennedy Space Center, where New Horizons was to launch. Those giant reptiles lend a bit of a menacing air to the causeways leading to NASA's launch pads, but they're also a metaphor for problems. "You know, if you have one alligator in the water and you know where it is, you can keep an eye on it and maybe outrun it," Todd explained. "But if you've got a whole bunch of them, man, it's hard to fight them all off." Alan:

> It was clear that Todd's analysis was correct—that we had sort of normalized ourselves to these kind of problems because the project had long been a NASA stepchild that we had kind of forced on the agency when it didn't want to do Pluto, when the "Clydesdales" at NASA had left us to fend for ourselves while the project's funding was uncertain. And even after the political and funding battles were over, we'd never had the kind of "we've got your back" support that NASA normally gives its projects. Now Todd had made it clear to us, and then to NASA Headquarters, that New Horizons had five or six serious problems, any one of which could sink the project schedule, unless NASA stepped in to help.

So Todd went to work to remedy the situation. Todd:

> We put a little team together and went systematically through each one of those problems. For each one, we ranked how severe

its risk was and made a detailed plan to propose to NASA Headquarters for how we would resolve it.

From his vantage point as liaison between the New Horizons project and NASA leadership, Todd could see clearly that the success of
the project was endangered by its history of having a lack of full support from NASA Headquarters. Todd:

> When we started, New Horizons just wasn't getting the atten-
> tion it needed from the Agency. I told NASA Headquarters,
> "These guys are in a crisis. They're working as hard as they can,
> but they cannot get from A to B unless you give them more re-
> sources and some serious backing. They're not going to make it.
> And when they don't succeed and this mission fails to launch or
> misses its Jupiter window, it's going to be a huge black eye to
> NASA. New Horizons and the exploration of Pluto is a highly
> visible project, and when you can't get it to the launch pad on
> time, it's going to be viewed as NASA's failure."

Alan was impressed with how quickly and completely Todd got
NASA's "Clydesdales" to start pulling for New Horizons. Alan:

> Todd really turned it around. He single-handedly got NASA
> Headquarters on the bus and pulling for us, helping us, taking
> ownership to make New Horizons succeed. With his help, and
> for the first time, I felt people at NASA were helping us. Previ-
> ously, due to all the bruising approval and funding battles to get
> New Horizons started, I had felt the attitude at NASA Head-
> quarters was more or less "You shoved this mission down our
> throats—good luck pal; we hope you make it." Simply put, there
> had always been either some passive-aggressive behavior going
> on, or maybe just benign neglect, but either way, that kept us
> from being able to solve many of our toughest management chal-
> lenges. Todd's biggest impact was probably that he got NASA to
> really change, to help us, and it made all the difference: after that

I really felt like we were being lifted on people's shoulders and carried forward to the launch pad.

With that help and a year and a half's hard work, by the late summer of 2005, New Horizons had completed construction. In addition, all its scientific instruments—even Ralph—were also completed, the project had garnered enough nuclear fuel to make the mission feasible, and even nuclear launch approval was on track to complete in time.

It had been an exhausting eighteen months since Todd May had arrived on the scene, ignorant in the details of robotic solar system exploration. But Todd's management mojo outweighed his lack of experience. Looking back, without his help, New Horizons almost certainly would not have made it.

There's no doubt, Todd May became a hero of New Horizons and a savior of the exploration of Pluto.

7

BRINGING IT ALL TOGETHER

PLUTONIUM POWER CHALLENGES

As we said in chapter 6, spacecraft that fly far from the Sun can't be powered by solar cells the way most spacecraft are. Instead, they use nuclear batteries called RTGs, fueled by plutonium.

As we also described in chapter 6, any rocket launching with plutonium on board is subject to a stringent gauntlet of governmental safety reviews and environmental approvals that take years to complete. The challenge in this for New Horizons was that while most nuclear missions had taken eight to ten years to navigate this complex regulatory process, the Pluto launch schedule only allowed four years to complete the same diligent work. Because there wasn't the possibility of skipping any of the nuclear launch approval process, the New Horizons team realized at the outset that they could only make the 2006 launch window if they made this complex process a special area of focus and put an experienced and skillful executive in charge of it. As we also noted in chapter 6, APL tapped Glen Fountain for this at the project's outset in 2001, three years before he became the project manager for all of New Horizons.

One of Glen's first tasks in this was to build the nuclear launch approval "data book." This thick set of documents is a compilation of all the information about the launch vehicle and spacecraft needed to perform the detailed safety, risk, and environmental impact analyses that go into achieving launch approval. The data book describes all the environments that the spacecraft would experience under different ground and launch accident scenarios, and how each could affect the integrity of its nuclear power source—the RTG. In short, the data book provided detailed answers and supporting analyses to determine the chance of a release of radioactive material, and if that occurred, how much material would be released over what area, and what the potential health effects would be. This in turn fed into a meticulous assessment of the probability that anyone would die due to radiation exposure for the different possible failure scenarios. Glen's role—bringing the data book together and then orchestrating the entire nuclear launch approval process, right up to the final approval in the White House itself, was perhaps the toughest job on the entire project. Alan:

> Glen is self-effacing about his role, but he literally made space-flight history by getting New Horizons through this labyrinth in time, allowing us to make the 2006 launch window. What Glen did was legendary.

In addition to navigating the tortuous regulatory maze, the actual physical production of the RTG nuclear power supply was itself a complex and laborious process, which was performed for NASA by the Department of Energy (DOE). Just as DOE had for previous nuclear-powered deep space missions like Voyager, Galileo, and Cassini, its job was organized in several efforts. For one, DOE's RTG contractor Lockheed Martin prepared the RTG for flight on New Horizons. Separately, DOE's Los Alamos National Laboratory produced the RTG's plutonium dioxide fuel, and formed it into ceramic pellets that would be loaded into the RTG. The RTG and the plutonium dioxide then met for testing at DOE's Idaho National Lab, a fortified defense lab replete with tanks, barbed wire, control towers, and heavily armed guards who follow visitors everywhere, even to the bathroom. Once the RTG was tested, it was shipped to Florida in a manner befitting a spy story. DOE and NASA sent it cross-country to Cape Kennedy in a covert but heavily armed convoy; in fact, DOE sent multiple convoys to the Cape—all but one containing decoy RTGs—to make it even harder for anyone trying to sabotage or pilfer the plutonium.

Everything the DOE did to prepare the RTG and fuel worked like clockwork, with one very nearly project-killing exception. Its Los Alamos National Lab, which is also involved in the production of nuclear weapons, was forced to halt all operations due to a security breach. The investigation lasted months while New Horizons waited, in agony, waiting for the resumption of its must-have nuclear fuel: not enough fuel—no Pluto mission. Finally, once the security investigation was completed Los Alamos was reopened, with barely enough time left for complete plutonium production. But then, in a new nightmare turn of events, Los Alamos was shut down again. This time it was because an accident elsewhere in the lab led to a safety issue—unrelated to New Horizons—but again putting the mission in jeopardy as it became clear that New Horizons was not going to get a fully fueled generator in time for launch. Alan:

> Our engineering team couldn't redesign the spacecraft to oper-
> ate on less power—it was just too late for that. Instead they started
> to look at ways to run the mission on less power, taking more
> risks, collecting less data, running fewer instruments at once,
> and so forth. We really were facing an existential crisis: if we
> couldn't figure out how to operate the spacecraft with a lower
> power level, there would be no exploration of Pluto, so we looked
> at every possible option.

They found that by cleverly operating the spacecraft, the mission could accomplish all its goals with as little as 190 watts at Pluto, about 35 watts less than originally designed for. That, and some spare pluto-nium fuel found from past NASA projects did the trick, because in the end the total plutonium DOE would deliver for the RTG was enough to produce just 201 watts at flyby. It was as close a call with fuel production as any RTG mission had ever had, but New Horizons survived it.

TESTING, TESTING . . .

By the spring of 2005, its Atlas rocket was being assembled, and the mission's plutonium problems were solved. Back at APL, all the various spacecraft subsystems and the seven scientific instruments were being mounted aboard New Horizons. After each was bolted onto the spacecraft, it was electrically tested to verify its functions. Then, once the entire spacecraft was assembled and functioning, the entire flight system was put through a battery of tests at APL—from launch vibration testing, to launch acoustics testing, to operating it from the New Horizons mission control.

APL then took the spacecraft—roughly the size and shape of a baby grand piano, south to NASA's Goddard Space Flight Center in Greenbelt, Maryland, to begin several months of testing. For this, New Horizons was put into a thermal vacuum chamber, where the air gets pumped out to simulate the vacuum of space, then it was repeatedly heated up and chilled to simulate conditions it would encounter in space. This was all to make sure that the spacecraft's systems, both primary and backup, would function as planned in flight, and to weed out any weak parts that might fail under these conditions.

And the testing did what it was supposed to do: it revealed problems so that they could be fixed—after all, there would be no fixing things once New Horizons was launched and on its way.

Most systems aboard the spacecraft passed their tests without problems. But a main computer failed and had to be replaced. And the Inertial Measurement Units (IMUs)—the gyros responsible for determining where New Horizons is pointed—leaked when subjected to vacuum. It took three rounds of replacements, racing against the clock with every replacement cycle, to find IMUs that would not leak. Changes in the thermal blankets were also needed as testing refined how they performed. And of course many software bugs were found and squashed.

Finally, after months of grueling tests in space-simulation chambers, parts replacements, and software fixes, New Horizons was able to pass all its tests with flying colors, and was certified as ready to ship to Florida for its launch from Cape Canaveral.

FOUR'S COMPANY

Something else was going on for New Horizons during the spring and summer of 2005 besides spacecraft testing, a wonderful scientific discovery. That story largely revolves around Hal Weaver.

In addition to serving as the mission project scientist, itself a demanding engineering-science job, Hal was also leading part of a long-standing effort that Alan had begun years before to search for additional moons of Pluto. Many on the New Horizons science team suspected that, in addition to Pluto's double planet partner Charon, the binary pair might be accompanied by some smaller moons that had not yet been detected. If so, knowing about them would be important for the detailed flyby planning, because if there were more moons, the team would need to know about them to choreograph observations of them.

By far the best tool for searching for small moons around Pluto was the Hubble Space Telescope, but it's never been easy to get observing time on the Hubble. The selection committees that allocate Hubble observing time typically receive seven to ten times as many proposals to use the telescope as there is available time. Twice during the development of New Horizons, Hal, along with science-team member John Spencer, Alan, and others, had proposed to use Hubble to search for faint satellites of Pluto. But both times they had been turned down.

It probably didn't help that the New Horizons team had also been searching for years using some of the best ground-based telescopes on Earth, and hadn't found anything. These negative results may have made a Hubble search seem a long shot that wouldn't be worth it. But countering this were computer calculations that Alan had made in the 1990s with his former postdoc, Hal Levison, showing that the Pluto-Charon system held many stable orbits for small moons to hide in, even moons up to one hundred kilometers in diameter.

In 2004, Weaver and others proposed to search with Hubble for the third time, and were turned down once again. Weaver was really surprised at this, thinking that they had so pared back their request—asking for just three hours of time on Hubble—that logic would be in their favor and their proposal would be approved. After all, the portent of such a discovery would be huge—both for understanding the Pluto system and for planning the Pluto flyby. But alas, even that had not done the trick.

Then, in the late summer of 2004, something very unusual happened: one of the major instruments on Hubble shorted out, ending its life. Since none of the observations with that instrument could then be conducted, the Hubble project looked around to fill the newly opened observing time. Hal remembers: "Suddenly, I got a call, and it was Christmas in summertime: we'd been awarded those three hours of observing time on Hubble after all. We'd get to make the search!" Naturally, though, that good news came with the usual New Horizons dose of delayed gratification—the Hubble observations would not be possible until May of 2005, due to scheduling constraints.

While they waited, Hal led the detailed planning for the Pluto satellite search observations. Then, once they got the data, he and one of Alan's postdocs, Andrew Steffl, began careful analysis to see if they could find anything faint orbiting Pluto. Hal Weaver is a soft-spoken, even-keeled guy. But Hal couldn't contain himself when the Hubble data revealed not one, but *two* additional moons of Pluto. Within days, Steffl, not knowing of Hal's discovery, found the same two new moons.

What a discovery! The Pluto system wasn't a binary after all, it was a quadruple, and New Horizons was going to be able to study not just one moon, Charon, but now three!

Having discovered two new moons around Pluto, the team got to nominate names for them. The way this works in planetary science is that the discoverers propose names that are used unofficially until the International Astronomical Union (IAU) signs off on the official ones. Because the IAU approval process is long and requires detailed naming proposals, the team adopted lighthearted provisional names to use while searching for, selecting, and then proposing what would become the official names that would go in the science textbooks. The temporary nicknames: Boulder and Baltimore, because almost everyone involved in their discovery was from one of those cities.

When the team later got around to naming the new moons for real, Alan wanted names that made sense with the mythology of Pluto but that also honored the tradition of how Pluto itself was named. Recall that back in 1930, during the effort to find a name for Clyde Tombaugh's new planet, Percival Lowell's widow wanted to call the new planet "Percival" or "Lowell" to honor her husband for initiating the search for the new world. Eventually, when eleven-year-old Venetia Burney suggested "Pluto," the scientists at Lowell Observatory liked it not just because of the underworld mythology surrounding the god Pluto, but also because it began with a *P* and an *L*, which could also be used to honor Percival Lowell.

For Pluto's two new moons, Alan, Hal, and the entire discovery team settled on the names "Nix" and "Hydra." In mythology Nix is the Greek goddess of darkness, and the mother of Charon. Hydra is a nine-headed underworld serpent (appropriate for a moon of the ninth planet). Nix and Hydra are fine names from ancient mythology, and they dovetailed well with the underworld theme from which Pluto and Charon had been named, but there was something more in these particular names that clinched them: like the *P* and *L* that honored Percival Lowell in the name "Pluto," Nix and Hydra began with an *N* and an *H*, and therefore could also be used to honor New Horizons, which had provided the reason to search for them in the first place.

A NIGHT FLIGHT TO THE CAPE

When all the spacecraft environmental testing at Goddard Space Flight Center was complete, it was time to ship New Horizons to

Florida for launch. The project had the choice of flying the spacecraft down on a military cargo transport, or driving it over one thousand miles in an environmentally controlled truck, surrounded by a convoy of support vehicles. The team decided flying would be safer. So, late on the night of September 24, the world's first and only spacecraft charged with the exploration of Pluto and the Kuiper Belt was flown from Andrews Air Force Base outside Washington, DC, to NASA's Kennedy Space Center launch site at Cape Canaveral. Alan and Glen and almost twenty spacecraft engineers and technicians who would be spending the next several months in Florida preparing their bird for launch flew down with it. Alan has vivid memories of that night flight to Florida:

> I remember flying down the eastern seaboard in that big Air National Guard C-17, thinking, "The next time the spacecraft is at this altitude it's going to be 'hauling the mail' spaceward on its Atlas, headed for orbit."
>
> From the panoramic windows of the C-17 cockpit, where the crew let me ride, I remember seeing the lights of cities and shorelines painting the entire eastern seaboard, and then eventually Florida ahead. As we began the landing approach, you could see the launch complex and NASA's huge Vehicle Assembly Building and the three-mile-long shuttle landing strip that we were aiming for.
>
> Once we landed, we taxied over to where various NASA people were waiting to receive New Horizons with an environmentally controlled truck. The truck would take New Horizons to its Florida clean-room home, where it would be prepared for launch as the team made final tests, fueling operations, and spin balancing.
>
> The air inside the C-17 was of course air-conditioned but outside at the Cape, even in late September at two in the morning, the Florida air was hot and muggy. When the back cargo door of the C-17 opened, the cool aircraft air met the dense, warm Florida air, and created an instant, surreal fog, billowing out the

back of the airplane. No one could have designed a more dramatic effect, even if this had been in a movie.

After the spacecraft was taken out of the C-17 and placed aboard its truck, Alan went back to grab his things and then descended the crew stairs. NASA's Chuck Tatro, the New Horizons launch-site manager, was there at the bottom of the stairs. As Alan stepped onto the tarmac, Chuck put his hand out to shake Alan's and said, "Dr. Stern, welcome to the launch site." The words hit Alan like a ton of bricks. Alan:

> After all these years, from 1989 to 2005, we really, finally had a Pluto spacecraft at its launch site. We were really about to fly across the solar system and explore the farthest worlds in history. The reality of the impending launch and decade-long flight across the solar system hit me when Chuck said, "It literally sent a shiver up my spine!"

8

A PRAYER BEFORE YOU GO

LUCKY 13

With their spacecraft at the Cape, the New Horizons contingent sent to Florida began a frenzied ten-week schedule of final spacecraft testing and launch preparations. Given the intensive work schedule, Alan and Glen took temporary apartments in Cape Canaveral for that whole period, as did others from APL and SwRI.

All the activity at the Cape was leading up to their single, three-week launch window coming up in January 2006, when Earth, Jupiter, and Pluto, all moving along in their orbits, would be arranged in just the right way to allow the Atlas to put New Horizons on its fast-paced, 9.5-year trajectory to Pluto. If they missed that twenty-one-day window in January, they would have to face a 2007 launch on the much riskier, fourteen-year-long journey that had no Jupiter flyby.

In mid-December, as spacecraft launch preparations were winding down and plans were being made to put New Horizons atop its Atlas launcher, Chuck Tatro came to Alan with a request. "We're on target to launch on the very first day of the launch window," he said, "but we want to give the team a few days off at Christmas and a couple more days at New Years. So we think you should give up five days of the launch window for that." Knowing the risks of not launching by January, and hence the risks of this decision, Alan asked to see the statistics on previous Atlas launches and how often they flew within what would now be only a 16-day window. Alan recalls the conversation:

> Tatro looked me in the eye and said, "I know this isn't an easy call for you, but we are going to make the launch window, Dr. Stern. We know how to do this, and the risk of giving up these five days is low. In fact, we think the risk of not giving the time up, of not giving the launch crew rest and the morale boost of time with their families at the holidays, is greater than letting the five days go."
>
> I knew how hard the launch team had been working, and I knew that the launch record showed the Atlas team rarely took more than a week, even with weather delays, to get launched once the rocket was ready to roll to the pad. So I made my decision and agreed to give the team time away for the holidays.

Done deal. But just as the team came back from those holidays at the first of the new year, now barely two weeks from the opening of the launch window, Tatro came to Alan again, with a detailed schedule for the dozens of steps still remaining to prepare New Horizons and its launcher for its countdown. Alan was familiar with the task flow, but now it was laid out on a calendar with actual dates for every step across the first half of January.

Tatro told Alan, "There's one thing we want to ask you about. If you look at this schedule, the day that we fuel the spacecraft with plutonium and it becomes electrically alive in its final flight configuration is Friday the 13th. This may sound a little silly, but does that make you uncomfortable?" Alan:

> They literally said, "We can fuel the RTG a day later, on Saturday the 14th, if you prefer, and we'll even pay the overtime for it. We just don't want you worried at launch or as you fly it to Pluto that the spacecraft went live on Friday the 13th." The first thing

that came to my mind was a boyhood memory of Apollo 13, and how some people then had thought they should never have given the mission the number 13, or launched it at 13:13 on the clock in Houston.

But then I reminded myself: I'm a scientist. Superstitions about Friday the 13th are completely irrational. So I decided that fueling on the 13th was preferable to giving up another precious day of launch window and that instead we would make Friday the 13th a rallying cry for the project. "From now on," I told myself, "every Friday the 13th we're going to celebrate the day as both the birth of New Horizons and a victory for rational thinking." I looked over at Chuck and said, "Oh, to hell with Friday the 13th. Fuel that bastard. Then let's light up that Atlas of yours and go fly!"

"BALLSIEST THING I'D EVER SEEN"

The final authorization to roll any NASA mission to its launch pad depends upon a launch approval document called a COFR: the Certificate of Flight Readiness. All the key NASA stakeholders for the flight have to sign it, certifying that the spacecraft and the launch vehicle and the spacecraft and the mission control and tracking network—all the elements of the mission—are ready to fly. Glen would sign for APL, and Alan would sign it as mission PI. About a dozen other key managers from Lockheed, Boeing, the Department of Energy, and NASA would also sign.

The signing of a COFR isn't just a ceremony. It's the final step in a long, careful, and technically complex launch-readiness process NASA conducts to make sure that every element of the project is ready. Literally thousands of items have to get checked and verified that they are ready in order for the various signatories to be authorized to sign the COFR.

When the New Horizons COFR authorization process began in the late summer of 2005, there was one issue about the launch vehicle that reared its head. It stemmed from an incident the previous summer at the Atlas V factory. Lockheed Martin had been testing a non-flight liquid-oxygen tank to verify it would hold pressure even beyond its flight design limit. To do this, the Lockheed team deliberately over-pressurized the tank to see if it would hold. But the tank burst, leading to a massive engineering investigation to determine why the proof tank failed.

The investigation lasted for months, examining every aspect of the proof tank's design, materials, manufacturing history, and handling. No stone was too small to turn over, right down to looking at the microscopic structure of the materials the tank was made of, testing hundreds of samples of that material, analyzing their strengths, searching for any weaknesses to explain the burst that shouldn't have been.

The investigation stretched on for the remainder of 2005. Of course, it had nothing specifically to do with the Atlas meant for New Horizons—its tank had passed every test—but with a plutonium-filled RTG aboard, the proof tank failure meant that New Horizons could not launch unless the investigation proved that cause had no relation to the materials, parts, or assembly processes of the Atlas that would fly New Horizons.

The issue came to a head in a NASA Headquarters meeting the first week of 2006, barely a week before the launch window opened. Called a Program Management Council (PMC), this high-stakes meeting had more than a hundred executives, managers, and technical experts in the room when the final decision was to be made. Ultimately, it would be up to then NASA administrator Mike Griffin—a brilliant, savvy, but then almost brand-new NASA boss—to make the call.

Much was riding on Griffin's decision. If something went wrong during launch, it would doom not only New Horizons and the possibility of exploring Pluto, but it might well make it almost impossible to launch future nuclear missions in general. An accident on New Horizons could, in effect, doom the future exploration of the outer solar system.

PMC meetings are normally for official NASA personnel only. But Alan felt that as PI of the mission he should be there to hear the

arguments and to weigh in himself—to give his own "go" or "no go"—
so he appealed directly to Griffin and was given permission to attend.

In that PMC, some made the case why New Horizons should launch,
others made the counterargument. There were technical presentations
and counter-presentations that went on, literally, for hours. The main
case, however, was prosecuted by the chief engineer of NASA's Kennedy
Space Center, James Wood. Wood, a bespectacled, sure-footed, mid-
career rocket man, was known for doing his homework and checking
it twice (maybe even three times). Wood made a detailed case that the
proof tank anomaly was completely unrelated to the tank New Horizons was flying, and he recommended launch. Griffin and his senior
NASA Headquarters lieutenants asked dozens of questions, probing
every aspect of the case Wood made. When Wood and the other engineers had finished their presentations, and Griffin and his staff had
finished asking all their questions, Alan stood up. From what he'd
heard, both as a former aerospace engineer himself and as the man
who had to face the consequences of whatever outcome the meeting
yielded, he had concluded from Wood's presentation that New Horizons was not at risk to fly, but that it was definitely at risk if the launch
was delayed to 2007 while more studies of the tank rupture took place.
Convinced of the logic of Wood's case, Alan addressed the room:

> I told them, "I just want to say, first, that for those of you who
> don't know me, I have been involved in almost a dozen NASA
> mission launch decisions, so this isn't my first rodeo. I also want
> you to know just how much is riding on this decision for us to
> launch this month." Then I explained why, if we didn't make the
> January Jupiter launch window, there wouldn't be another like it
> for a decade, and that our only option would be to fly in 2007 on
> a long, slow 14-year flight to Pluto without a Jupiter gravity as-
> sist. I also pointed out that if we couldn't launch in 2007, even
> the Atlas V 551 couldn't get us to Pluto if it launched in 2008 or
> 2009, in fact not until 2014 or so. Then I carefully explained the
> much higher risk to the spacecraft of a 14-year flight time com-
> pared to a nine-year flight time, and I also noted the greater

expense of both the yearlong delay in launch and the four-year delay in arrival time. Then I gave the reasons why we wanted to get to Pluto before the atmosphere froze out, which also argued for an earlier launch and arrival, and how if we arrived later there would also be less of the surface in sunlight to map. Finally, I closed, making it personal. I said, "I have put 17 years into this project, stretching back to 1989. It's the NASA administrator's decision, but as the mission PI, and as someone with a lot riding on this launch, I want you to know that based on the data, I have no qualms whatsoever about launching this Atlas as is." Then I sat down. My case was made. A couple of people rubbed me on the shoulder in that kind of way that means "Nice job, but I hope you can handle whatever Griffin decides to do."

After Alan spoke, Griffin polled all of the major heads of NASA for their recommendations on whether the Atlas V tank issue was sufficiently settled for New Horizons to launch. Many voted to launch, but some voted against it, saying they had no solid reason to doubt the New Horizons Atlas, but they just didn't want to take any chances. The seasoned head of NASA Launch Services, Steve Francois, responsible for all NASA rocket launches, voted to fly. So did NASA's chief engineer, Rex Geveden, as did the head of planetary exploration at NASA, Andy Dantzler. But the head of NASA's Office of Safety and Mission Assurance, Bryan O'Connor, a former space shuttle commander voted not to launch.

Next, Mary Cleave, another former astronaut and the head of all science missions at NASA, voted no. Alan: "I was thinking to myself, 'Griffin can overrule these guys, but if he does, he'll now be on record as going against both the head of safety and the head of science. If this launch doesn't succeed, no matter what the reason, he'll probably lose his job as NASA administrator and his career.'"

Hal Weaver, who was also in the meeting, remembers feeling incredibly discouraged. Hal recalls, "I was getting pretty depressed. Of course you might expect the safety and mission assurance officer to vote it down, because his head will roll if something goes wrong if he had voted to launch. But the head of all science missions?"

The last to vote was Griffin's most trusted lieutenant, Bill Gerstenmaier, responsible for all NASA launches. Gerstenmaier voted to launch, explaining calmly and carefully why he thought that the proof tank–burst issue had been studied and analyzed thoroughly and that the analysis, presented by Wood, had clearly exonerated any risk for the New Horizons launch. He said that launching was the rational choice.

Then it was Griffin's turn. Speaking as the head of NASA, he stood up and gave a long closing argument to the room. The room was totally silent. Everyone knew that, with split opinions within Griffin's executive team, it was up to him to make the final decision.

Griffin reiterated that the burst of the proof tank had nothing specifically to do with the New Horizons launcher and that the flight safety record of the Atlas was perfect. He praised the analysis of the proof tank failure and laid out his own rationale for why the proof tank was unlikely to affect the rocket New Horizons was sitting on, which was not expected to see any tank pressures even close to the burst level. Griffin then pointed out that the probability of a launch failure was 2–3 percent on any launch, and that the risk associated with the tank was far lower, so the net effect of the oxygen-tank issue was only a small factor in the risk calculation. Then he reminded the room that every RTG was built to survive even a catastrophic launch vehicle accident and that, in fact, long ago just such an accident had occurred and RTGs had demonstrated that they do survive. Griffin's logic was cold, rational, quantitative, and impeccable. It was devoid of emotion, as he carefully and thoroughly reviewed the facts, concluding that the risks related to the proof tank failure were demonstrably minimal and that the risks of waiting were almost certainly greater.

Griffin concluded the meeting by overruling Cleave and O'Connor, and declaring that as NASA's chief, he had concluded that the vehicle was safe to launch. He signed the COFR right there, in front of everyone and walked out of the room. Alan:

It was the ballsiest thing I'd ever seen in my twenty-plus-yearlong career working NASA missions. It was literally like something in the movies. The NASA administrator just put his job on the line to overrule the head of safety and the head of science missions to give New Horizons permission to launch. Hal and I just looked at one another not quite believing the drama we'd just seen. Griffin proved both his spine and his spunk that day, and he became another hero of New Horizons.

SAY A LITTLE PRAYER

As launch week approached, people started to gather around the Cape by the hundreds, and then by the thousands. In addition to the engineers, managers, launch staff, scientists, and others directly involved with the mission, there was a growing army of journalists, documentarians, students, and teachers, as well as thousands of space fans and curious onlookers coming to view what promised to be a history-making launch. There wasn't a hotel vacancy to be found within an hour's drive of the Kennedy Space Center.

Among those in attendance were many of the original members of the Pluto Underground, who had been working since 1989 to get a mission to Pluto. Also there were other members of the planetary science community, people from the engineering teams, executives from every major partner in the project, and Mike Griffin. Nothing like this—the launch of a first reconnaissance mission to a new planet—had occurred since 1977, when the twin Voyagers launched to explore the giant planets.

A few days before launch, members of the New Horizons science team gathered for a final, daylong prelaunch meeting. Alan spoke to this group, reminding them of just how far they had come, and how after all those battles over all those nearly seventeen years, they had succeeded. Now they were gathered together on the eve of launch, with their extraordinary spacecraft mounted atop a rocket the size of a city skyscraper, ready to leave Earth, forever.

They had come very far indeed, but after Alan spoke, it struck him that everything they hoped to learn, everything they had worked for, depended on a successful launch, and then the almost ten-year, three-billion-mile journey ahead. Everything, really, was still ahead of them.

The next night, Alan went out to the launch pad with Todd May and Rex Geveden. Almost no one else was there. But there was New Horizons, atop its twenty-two-story-tall behemoth launcher called Atlas. The image of that rocket burned itself into Alan's memory. He knew it would only be there briefly, for a few more days, and then it would fly. It would be gone, and whether it succeeded or failed, he would never see it again.

The sea breeze washed over him; he could smell the salty, coastal Cape Canaveral air, a familiar memory from other launches he'd been involved in. He looked up at the rocket and quietly spoke to it: "Make us proud." Then he turned and walked back to his car.

The next morning they would count down to launch to Pluto.

9

———————

GOING SUPERSONIC

As the opening of the launch window neared, out on Cape Canaveral's Launch Complex 41, New Horizons sat atop its giant Atlas rocket, powered up and ready to fly. The "stack," as the rocket guys called it, was impressive—over two hundred feet tall. At the top, little New Horizons was attached to its STAR 48 solid rocket motor, both co-cooned together inside the cavernous Atlas nose-cone fairing built for school bus–size spacecraft.

Alongside this rocket stack was its gantry, which is a utility tower nearly as high as the rocket itself, with various conduits, cables, and umbilicals attached, attending to the rocket's needs for power, fuel, cooling, and communications until the moment of launch. The launch complex itself is adjacent to the beach, and miles away from almost anything else.

Much of Kennedy Space Center is kept as a wildlife sanctuary; in fact, less than 10 percent of the land is developed. The raw Florida marsh there is an excellent place for spotting egrets, ospreys, eagles, herons, and of course actual alligators. Across the marshes, dunes, and lagoons of this rich ecosystem are many iconic NASA facilities known from pictures and documentaries, including the enormous 525-foot-high Vehicle Assembly Building, a dozen or more other launch pads, the astronaut quarters, the Shuttle landing runway, and a sprawling visitor complex.

No doubt, there was something that drew people to this particular launch—a sense of something epochal, a passing of the torch from Voyager to a new generation of explorers who had been inspired by Voyager. You could feel it; it was in the air, now it was a new generation's chance to explore never-before-seen worlds.

New Horizons touted the slogan "the first mission to the last planet," and it did promise in many ways to be the "last major first" in the opening era of planetary reconnaissance of our solar system. It was also the first of a new class of NASA planetary missions—a series of $1-billion-class, competed, and scientist-led interplanetary exploration missions dubbed the "New Frontiers Program."

For those working on the mission, there was a keen awareness of both the decade-long flight time ahead, as well as the nearly two-decades-long struggle that had already passed just to get to this moment. For all these reasons, there was a sense that this was an important historical moment, and almost anyone who was anyone in space exploration came to it—from astronauts to politicians to planetary scientists from around the globe, to space news media and politicians.

By a weird, cosmic coincidence, the launch happened to be taking place near the anniversary of Clyde Tombaugh's passing, on January 17, 1997. This made the occasion particularly touching.

But there was another reason why this was an emotional event, particularly for the Tombaugh family. Unbeknownst to the general public, a bit of Clyde's ashes had been tucked away on board New Horizons. The idea to do so had originally been hatched by Rob Staehle, back in the 1990s, when JPL was studying the Pluto Fast Flyby mission. Rob had proposed the idea to Clyde, with whom he had become friends, and Clyde accepted. So, in early 2005, when it was beginning to look like the launch of New Horizons would soon be a reality, Alan raised this delicate topic with Clyde's widow, Patsy, and daughter, Annette, asking them if they knew of Clyde's conversation with Staehle, and if they had in fact saved some of his ashes to go to Pluto. Their response was an immediate and enthusiastic yes to both questions. They told Alan that Clyde had wanted this very badly. So Alan asked his spacecraft engineers how one would actually do this, how they could mount a small container on the bird, because in spaceflight even something sentimental needed to be engineered. The engineers designed a small container that they would affix to a spacecraft wall and use to replace a small balance weight.

One day in mid-2005, Alan received a small packet of ashes from Clyde's family, which he physically carried out to APL in his briefcase and handed to the engineers to place aboard inside the container. On the outside of the container was a tiny plaque, inscribed with words Alan wrote: INTERRED HEREIN ARE REMAINS OF AMERICAN CLYDE W. TOMBAUGH, DISCOVERER OF PLUTO AND THE SOLAR SYSTEM'S 'THIRD ZONE.' ADELLE AND MURON'S BOY, PATRICIA'S HUSBAND, AN-NETTE AND ALDEN'S FATHER, ASTRONOMER, TEACHER, PUNSTER, AND FRIEND: CLYDE W. TOMBAUGH (1906–1997).

Think about that for a minute: seventy years earlier, photons of light from the Sun had reflected off Pluto, traveled for four hours and over all those billions of miles to Earth, and passed through a tele-scope in Flagstaff, Arizona. Those photons created a tiny dot in a plate of photographic emulsion that had caught young Clyde Tom-baugh's eye when he examined that image a few weeks later, revealing the existence of a new, faraway planet. Now some atoms that had been part of Clyde were going to make the journey to that faraway world and then continue on, outward, to leave our solar system for interstellar space and the galaxy beyond. Whatever you believe about life, death, consciousness, and fate, this was surely a unique and won-drous memorial, unlike any other in history.

COUNTING BACKWARD, TWICE

As planned, on Friday, January 13, 2006, the Department of Energy fueled the RTG aboard New Horizons. Replete with a phalanx of armed guards, NASA and DOE brought in its nuclear fuel. In the clean room, high up on the rocket at the level where the spacecraft was, there was a large, squarish hatch on the side of the nose cone, about five feet by five feet, used to provide access to the bird. In order to push the glowing-hot radioactive fuel into the RTG, specialized tools that could be operated from twenty feet away were used, safeguarding workers from the radioactivity. Then, from a distance, they installed the RTG's cap and tightened all its bolts to launch specifications. Once the RTG was fueled, its heat began producing electric power. From that moment on, New Horizons was alive, in the sense that it was producing its own power. And from that point on its systems were on and it was being operated from mission control back in Maryland, just as if it were in flight. In a very real sense, although New Horizons was still on Earth, its mission had begun.

The morning of Monday, January 16, was clear, cool, and sunny. The only weather concern was a forecast of high winds from a front that was blowing in over central Florida. Alan got up hours before dawn, answered email, did a ritual prelaunch run across the streets of Cocoa Beach, kissed his wife Carole goodbye, and headed in to the Atlas Spaceflight Operations Center, or ASOC. The ASOC is a large mission-control complex located just three miles from the launch pad. Alan, like more than one hundred others involved in the launch, took his place at a launch console, grabbed some coffee, and put on a communications headset. He had gone through these motions many times before in New Horizons launch rehearsals, but today was different. This was a live count, as evidenced by the ambulances outside the ASOC in case anyone had a heart attack, by the sprawling press contingent, and by the unusual presence of NASA administrator Mike Griffin in the control center.

Meanwhile, crowds of New Horizons team members, friends and family, planetary exploration fans, and the public, were all being bussed in to the various viewing areas around Cape Canaveral. A few dozen dignitaries and key visitors were ensconced at a VIP viewing site by the iconic Apollo/shuttle Vehicle Assembly Building, five miles west of the launch pad. Many of the science team members were gathered, with their families and friends, at a different viewing area with bleachers about five miles south of the launch site—as close as anyone in the open was allowed to get to the violence of launch. Others, by the thousands, had to take less prized positions farther away.

The science team members and their families could see the tall Atlas in the distance, surrounded by the four lightning towers of Launch Complex 41, across the wide watery expanse of Florida's Banana River. Through binoculars the rocket looked like a living, breathing thing, relentlessly venting liquid oxygen steam, in anticipation of its soon to be moment of glory or ruin. They could see that the main stage of the Atlas had turned from its normal metallic color to a bright white that matched the color of the nose cone on top, another sign that the Atlas had been fueled full of its cryogenic liquid oxygen and hydrogen, forming a layer of frost on the rocket's thin metallic skin.

After the busses dropped people off at the viewing sites, they were briefed about safety and shown the position of a nearby building where they could take shelter in the event of a bad launch accident. This warning didn't dampen the enthusiasm of the crowd, but it did remind them of the very real "playing for keeps" stakes.

The waterfront filled up with tripods and cameras and people staking out good spots to photograph and watch the launch. Kids ran and played and chased and tackled each other on the wide grassy expanse between the bleachers and the water. There was a giant digital clock displaying the official countdown time, and a set of loudspeakers carrying the voice of mission control. If all went smoothly, the clock would count down to four minutes before the launch, and then there would be a scheduled ten-minute "hold" while all the rocket and spacecraft systems were checked one last time. Then if everything was Go, the final four minutes of the countdown would proceed.

Based on the inexorable celestial mechanics of getting New Horizons on its trajectory to Jupiter, launch was scheduled to occur no earlier than 1:23 eastern time that afternoon. The same celestial mechanics dictated that the launch window would last less than two hours: if the bird was not in flight by then, they would have to stand down and launch another day.

The launch count proceeded normally until 1:17 P.M., barely six minutes from when the rocket should fly, but then a valve that did not seem to be opening properly and a rise in the low-altitude winds caused a delay to 1:45 P.M. At 1:40 P.M. the launch was delayed again to 2:10 P.M. The valve problem was fixed, but there was still concern about the winds. Then at 2:10, NASA announced that an issue had cropped up with a Deep Space Network (DSN) antenna station that needed to communicate with New Horizons after launch. That drove home how many things had to be performing perfectly at the same time, in so many different places, in order for everything to be Go for launch. It wasn't just the spacecraft and the rocket, or even the ground systems in Florida that had to be ready. There was also mission control at APL in Maryland, the launch safety equipment in Florida, and the entirety of the Deep Space Network of antennas around the world that also had to be ready, simultaneously.

Meanwhile, the wind began whipping up off the water, delaying launch yet again, pushing things uncomfortably close to the end of the day's launch window. The crowd began to wonder if they would even see a launch. After another delay to 2:50 P.M., NASA announced, "All launch elements report they are prepared to support launch today." Then more wind and another delay, to the very end of the launch window. Any further delay and the launch would be "scrubbed" for the day.

Finally, the countdown descended all the way to that four-minute hold mark. At that point, the launch director stopped to poll the responsible parties for each of the major systems on the spacecraft, the rocket, and the ground systems: "Go" or "No Go." Those listening to the public-affairs channel heard them each quickly give their leave, one after the other, giving their "Go."

Atlas? "Go." New Horizons? "Go." APL Mission control? "Go." A dozen more, then, finally. PI? Alan reported, "Go."

With each "Go" a little cheer erupted at the viewing sites. But then, at 2:59 P.M., the weather soured again. The announcement "We have a 'No Go' due to a red-line wind monitor," came over the loudspeakers. Surface winds were exceeding 33 knots at the launch pad, the limit for what the giant Atlas could correct for as it pushed its way past the launch tower. There was no more time, New Horizons would not launch that day. The exploration of Pluto would have to wait.

The next morning was January 17, the ninth anniversary of Clyde Tombaugh's passing. Weather forecasters gave a 40 percent chance of thunderstorms, but launch preparations were proceeding, so the crowds drove back out to Kennedy Space Center, ready to load onto buses that would take them back out to the launch viewing sites. Little did anyone know that over at the ASOC, a tense drama had been unfolding for hours.

Alan had arrived at the ASOC at 5:00 A.M., after another ritual, prelaunch run, just as he had the day before. That morning's run was filled with thoughts of both Tombaugh's passing and the many checklist items ahead to get the bird into flight.

Arriving at 5:00 A.M. may seem early for a launch scheduled for early afternoon, but the launch preparation procedure is long and time-consuming. Just as Alan arrived, he was informed that APL's power was down in Maryland. As it turned out, the same weather front that had blown through Florida the day before, raising those troublesome winds, had intensified and was now raging through Maryland. The storm had become so violent there overnight that it had knocked out the power. The New Horizons mission control was operating on only backup, generator power. Alan:

> I thought to myself, "Do I want to launch with our mission control on backup power? If New Horizons launches and needs mission control's help because of an anomaly as it reaches space, and that backup power to our mission control center fails, we'll have no way to help the spacecraft. We did not come this far to take an unnecessary risk like this at launch. If we were out of launch days, I might have to take this risk, but we still have two weeks of launch window ahead."

At the Mission Operations Center at APL in Maryland, Alice Bowman was at the center of mission control. An über-competent, fastidious, and calm engineer, Alice was (and still is) the New Horizons Mission Operations Manager (an acronym affectionately called MOM), meaning she was in charge of the team operating the spacecraft. Alice had been part of the project ever since the proposal effort back in 2001. She felt that she and her team were trained and prepared for anything the spacecraft could throw at them. But this? Alice:

> I got into APL about 5:30 A.M., which was closed for all except essential personnel because of the power outage. Of course, on launch day, my team was considered essential personnel. When I got to the control center, it was mostly dark. We had electricians frantically working to hard-wire the backup generator into the electrical panel. There were all these extension cords running around the floor because we were trying to figure out a way to combine different subsystems people around one machine. We only got everything rerouted about ten minutes before the opening of the launch window that afternoon.

The launch managers at APL and the Cape were confident that the backup power was rock-solid and ready to support the launch. New Horizons chief engineer Chris Hersman made the case that it would take two independent failures to get into a bad situation—one on the spacecraft and also a failure of the backup power generator at APL. Generally, the rest of the team was prepared to launch with the APL mission control on emergency generator power. But Alan was not, and he was resolute.

As Maryland's electric company worked to restore APL's primary power, the count continued. The Atlas was fueled, and again it turned that beautiful white from water in the thick Florida air freezing onto its icy skin. The spacecraft and third stage were prepared for launch. DSN antenna stations around the world checked out. Launch time approached, but APL was still on backup power when it was time for the final Go/No Go poll in the ASOC.

Launch director? "Go." New Horizons project manager? "Go." APL director? "Go." Alan recalls:

> As the launch director went around the communications loops polling everyone, over 20 mission managers all said "Go." But I just thought in my heart of hearts that if we launch and New Horizons gets into trouble and then our mission control goes down, I'll never forgive myself. "This," I thought, "is what it means to be the mission PI. The moment when rubber meets the road in a tough call."
>
> Over my headset, I heard the launch director ask me for my "Go" or "No Go." "PI?" Heads turned to look my way, because they knew I'd been lobbying to stand down all morning. But now the decision was mine. I said, "I'm not comfortable launching without two sources of power at New Horizons mission control. The PI is No Go for launch."

That was all it took. The PI was No Go. The launch was scrubbed. Thousands of visitors and hundreds of launch personnel would once again have to wait, as would the exploration of Pluto. Again, the rocket was drained of its cryogenic propellants, and the launch crowds dispersed to the area beaches, nature reserves, restaurants, beds, and bars.

Meanwhile, the launch team itself continued working, and looking at weather forecasts for the next few days. Alan asked APL to get a second generator in place, a second source of backup mission control power, so this couldn't happen again. APL agreed, reporting that it could have it in place by evening.

All of this *launchus interruptus* created an interesting dynamic. There was so much to do leading up to the day and the moment of launch, and with every attempt, all of it had to be repeated. Each time the launch team had to fuel, unfuel, refuel, and engage a massive amount of machinery and personnel—across the United States and across the world—to be geared up for the next launch attempt.

For those not directly involved in the launch—the bystanders, press, and team members there with families to witness their baby leave the planet—there was a *Groundhog Day* aspect to the repeated launch attempts. The various mission contractors threw hotel parties the night before each launch count, catering the exact same foods and drinks each time. Although the launch and mission teams were hard at work, there was a whole other life of parties going on for all the visitors. That, combined with understandable prelaunch anxiety, and a little sleep deprivation, made the entire repeating pattern seem a bit surreal.

IT SEEMED LIKE SCIENCE FICTION, BUT IT WASN'T

After taking a day off for the launch crew to take a break, on Thursday, January 19, it was time to try again. The morning was chilly but sunny, and there was almost no wind, just a scattered deck of low clouds. The feeling of the crowd was guardedly optimistic. Over at the ASOC, the mood was businesslike for the third countdown and launch attempt for New Horizons.

On this try, the drama became about the clouds and whether they would become broken enough to fly, so once again an important Go/No Go became that of the launch weather officer. Due to the changing positions of the Earth and Jupiter in their orbits, the launch window opened a little bit earlier each day. So on the 19th, the first attempt was set for 1:08 P.M. Eastern Time. The clouds came and went, resulting in a series of short delays as the Atlas team tried to find a time that would allow their rocket to dodge the clouds. Finally, a new "T–0" time was set for 2:00 P.M. The clock ticked down all the way down to T–4 minutes, and the mandatory ten-minute hold kicked in for all of the engineers and launch officers to make their final checks before the sprint to launch.

A group of old friends from the New Horizons science team had gathered at a viewing site, anxious for their baby to take flight after so many years of hard work proposing and building and planning for its journey to Pluto and the Kuiper Belt. Science team geology and geophysics team lead Jeff Moore was there with his daughter and his mom. Geologist Paul Schenk was with his husband, David; Leslie Young was with her husband, Paul. Bill McKinnon had his kids in tow. Carter Emmart and David Grinspoon, who had both worked closely with the team on public outreach, were also there, sharing in the anxious excitement. Planetary astronomer Henry Throop prowled with his camera, smiling and shooting pictures of everyone and everything, capturing what could, if they launched today, turn out to be a pivotal moment in their lives. They had all been staring at the Atlas, waiting for it to launch, for days now. But it had just been sitting there, steaming off liquid oxygen in the Sun. Even from miles away it looked enormous, like a prominent skyscraper built along the shore. In an odd way, it seemed permanent, and it was hard to imagine that something so dramatic was about to make something so large lift off and fly away.

A tense hush fell over the viewing site when the clock picked up after the hold at T–4 minutes. It was the first time New Horizons had gotten so close to launching.

Back in the ASOC, Alan was at his console. Everything was going exceptionally well, save one minor portent. Alan was taking careful notes in his project diary using the same pen that he'd used for all of the previous countdown attempts and all of the launch rehearsals and simulations. Suddenly, just minutes before launch, the pen ran out of ink. "What? Wait. Why now?" he thought. But the moment passed and he shook off the seeming omen.

Four minutes ticked down to three, then two, then one. In the final prelaunch poll Alan and all of the other launch managers gave their "Go's." And then, with his launch responsibilities complete, and just thirty seconds left in the countdown, Alan stood up, took his headset off, and ran as quickly as he could to a secret door he had found earlier, the only unlocked outside door in the ASOC.

Out at the viewing site the crowd held its collective breath as the countdown ticked away through its final seconds. The launch team's commentary could be heard over the public address system:

"*Third stage is Go.*"

"*Roger.*"

"*Minus twenty-five seconds.*"

"Status check."

"Go Atlas."

"Go Centaur."

"Go New Horizons."

"T minus eighteen . . ."

". . . fifteen seconds . . ."

". . . eleven seconds . . ."

At T–10 seconds the crowds began shouting the seconds along with the announcer: "four . . . three . . . two . . . one! . . ." A billow of smoke and steam shot out from the base of the rocket, and as the count dissolved to zero, a blindingly brilliant spot of light erupted at the base of the Atlas, which started to move. The light widened into an incandescent cone as the rocket accelerated and cleared the tall launch tower in under two seconds. New Horizons was in flight!

The launch announcer marked the moment, saying, "Liftoff of NASA's New Horizons spacecraft on a decade-long voyage to visit the planet Pluto and then beyond!"

The behemoth Atlas was accelerating furiously upward. The flames beneath it lengthened to about twice as long as the rocket was tall, and they were unbelievably bright: even miles away it hurt your eyes to look at it. But you couldn't look away: it was mesmerizing. A rocket was leaving on the most distant journey of exploration in the history of our species. It seemed like science fiction, but it wasn't!

For those there, the first few seconds of the launch had been an all-visual experience, because the sound from the event had yet to reach the crowds from the launch pad miles away. But then the rolling thunder of the Atlas swept across to the viewing sites. If you've been up close to the band at a really loud rock concert or under the thunder of a high-performance military air show, then you know something about what this is like, rattling your body with intense, low-frequency, staccato vibrations. Every frequency shook all at once, so that every cell of your body quivered and vibrated as the Atlas, carrying precious New Horizons, rose on a pillar of smoke higher than any mountain on Earth, and then arced, now supersonic, out over the Atlantic.

Back at the ASOC, outside on a balcony, Alan watched alone as his baby lit up and pierced skyward, straight into the blue. He knew it was against the rules to be outside this close. But after 17 years of work, he was not going to watch this on a monitor. No one else saw the launch from so close. As the sound washed over him, louder and louder, he repeated to himself, again and again, "Go, baby, go—make us proud!" Then, when the Atlas disappeared behind clouds as it was making its gravity turn eastward, Alan sprinted back inside to join the launch team in the control center. He put his headset on, and went back to business.

The launch team carefully monitored the rocket's systems and trajectory, measured its performance against plan, quietly ticking off successive milestones as the launch proceeded, step-by-step, spaceward.

After 105 seconds, the five massive solid rocket boosters of the Atlas first stage were jettisoned, having done their heavy lifting. Through binoculars you could see five little white needles tumbling and falling away as the Atlas continued to climb, its steady glow finally disappearing behind some high clouds. New Horizons was now out of sight, forever.

After three minutes, the rocket was so high that it was above all air resistance, and its nose cone was jettisoned, leaving New Horizons, at the point of Atlas's gigantic spear, out in the open. New Horizons was in space—the place for which it was built—for the very first time.

After four and a half minutes, the first stage of the Atlas had used up its fuel and was jettisoned. The Atlas's Centaur second stage then lit up and fired for almost five more minutes, accelerating itself, the third stage, and New Horizons to reach Earth orbit at eighteen thousand miles per hour—barely eight minutes after it left its launchpad.

Watching the monitors in the ASOC, as the Atlas was confirmed in orbit, Alan felt a hand pat his back and the friendly voice of NASA launch director Omar Baez, "Dr. Stern, welcome to space."

New Horizons raced across the Atlantic, and then across North Africa, more than one hundred miles above our planet. It was spending the first hour of what would now be an eternity in space.

Half a world away from the launch site, over the Middle East, the Centaur and New Horizons reached the mathematically calculated point at which the engines needed to fire again to boost the space-craft onto its trajectory toward Jupiter. In the ASOC, telemetry showed the Centaur ignite again, precisely on schedule. Ten minutes later, Centaur completed its job and separated from the third stage, with New Horizons atop it. The spacecraft and its third stage were now moving fast enough to escape Earth's gravity, yet not yet fast enough to reach Jupiter, and therefore Pluto. But 34 seconds after the Centaur separated, the Boeing third-stage STAR 48 rocket motor erupted on schedule. It only burned for 84 seconds, but it performed perfectly and accelerated New Horizons to 14 G's and to its designed terminal speed—faster than any spacecraft had ever been launched.

The Atlas, the Centaur, and the STAR 48 had all done their jobs. New Horizons was now screaming away from Earth, toward Jupiter.

For Alice Bowman, back at the MOC in Maryland, the moment of truth came just seconds later, when New Horizons turned on its transmitters and APL started receiving telemetry from the spacecraft. Alice recalls, "We cheered at the launch, but not as hard as we cheered when we got that telemetry, because that's when we knew that the spacecraft had survived launch and was doing fine. We had a mission, and that was when we popped the champagne!"

Within only a few more minutes, tracking data revealed that the launch was an absolute bull's-eye; the targeting was even better than prelaunch predictions. New Horizons was not only operating flawlessly, it was also right on its planned course.

When Alan heard that report over the comm loops, he threw that dry ink pen he'd worried about into a trash can. The ASOC erupted in cheers, hugs, handshakes, and high fives across its many control rooms.

They had done it! Against all the struggles, doubts, and naysayers of the past 17 years, a spaceship had left Earth that day on its way to explore the Pluto system. With it rode the hopes of its team and a larger scientific community for what discoveries it would make there, a decade hence, in the cold, cold reaches of the outer solar system. Also aboard New Horizons, now silently suspended in weightlessness, were some of Clyde Tombaugh's ashes, speeding on their way to the planet he'd discovered so long before.

A BONFIRE ON A BEACH

That night it felt like the whole Florida Space Coast became one big joyful party for New Horizons. People were honking horns on the streets. Strangers were stopping strangers just to shake their hands. Celebrations erupted everywhere.

The biggest post-launch party raged at a high-rise hotel on the Atlantic shore south of Cocoa Beach. By the time Alan arrived, delayed by press interviews back at the launch site, the party was already well under way, and the bar had been open for quite a while. The atmosphere was euphoric.

That revelry went on and on into the night. Eventually, an excited crowd gathered around the smoke and glow of a fire set in a big, barrel-shaped trash can on the beach behind the hotel. The people around it were clutching beers and martinis and mai tais and margaritas, a gaggle of nerds who had something special to celebrate. Alan:

> Someone brought me over to that bonfire and told me what was up. The Atlas team had a tradition of burning all the now-obsolete emergency contingency procedures books after a successful launch. The team gave me the honor of throwing the last of them into the flames, which I did—with gusto! That Atlas team bonfire was the coolest post-launch tradition I'd seen in a long while.

As the bonfire flames leaped above the beach and into the Florida night, New Horizons sped away from Earth at 36,000 miles per hour, so fast that it made it to the Moon's distance ten times faster than any Apollo mission had—just nine hours after launch, and at that point, it entered interplanetary space, its new home.

10

TO JUPITER AND THE OCEAN
OF SPACE BEYOND

LEARNING TO FLY

Following the excitement of launch, the crowds quickly dissipated. After a flurry of public exposure, with the launch of New Horizons featured in newspapers and television news reports and splashed across the covers of magazines, the press scattered as well. But the real work of spaceflight was just beginning for the small team of New Horizons engineers, scientists, and flight controllers charged with flying the mission across the solar system to reach Pluto.

New Horizons, having once been merely an idea, then an underdog proposal, then a canceled mission, and then a four-year mad dash to launch, was now what it was meant to be: a flight project.

Alice Bowman and her team dove immediately into the complex work of getting the spacecraft fully checked out and operating in space. First, they had to "de-spin" the spacecraft. When it separated from the STAR 48 third stage on January 19, it had been spinning like a dervish. This was by design, for gyroscopic stability during the solid rocket motor firing. Alan stayed in Florida for a day after the launch because he didn't want to be in the air and out of communication until their baby passed this critical spin-down, which it did. Then he caught a flight up the east coast to APL to spend three weeks with the operations team and "live with" the spacecraft for its early phases of checkout and the first course corrections to fine-tune the trajectory toward Jupiter. In theory, Alan could have headed home at this point and consulted during the daily operations ("ops") telecons, but he wanted the "high bandwidth" communication of being there at APL in case there were any problems that required complex resolutions.

During the first weeks of flight, each onboard system—communications, guidance, thermal control, propulsion, and all the others—was tested thoroughly. So were all of the backup systems. This was a painstaking process that involved dozens of test procedures being sent up to the spacecraft by radio. Each test was followed by the transmission ("downlink" in spacecraft-speak) of data from the test, which the APL engineering team pored over for the slightest sign of problems or unexpected behavior.

During these checkouts, New Horizons was already millions of miles away from Earth. Alice and her operations team were learning to fly New Horizons. Of course, they had practiced before launch, but now they were playing for keeps. So everything they did in those first weeks, every time the spacecraft was asked to do something new— turn on this system, turn on this backup system, maneuver a new way—felt to the team like it was playing with fire.

Their major concern, of course, was that New Horizons would point away from Earth or otherwise lose its ability to communicate with them, causing them to lose the spacecraft entirely. The history of spaceflight is replete with some tragic examples of this type of accident. NASA's Viking 1 Lander, the first successful Mars lander, had been lost six years after it touched down on the Martian surface, when a software update, meant to correct a battery-charging error, had accidentally included commands to redirect the communication dish. The errant commands pointed Viking 1's dish antenna at the ground, where it could no longer communicate with Earth, and just like that—the mission was over. Similarly, Russia's Mars probe called Phobos 1 was lost in 1986 when a single character was missing from a software upload. This minuscule mistake caused the spacecraft to deactivate its attitude-control thrusters; as a result, its solar panels could no longer track the Sun, and the batteries lost all their power. It was never recovered. And in one of the most painful and embarrassing failures in the history of spaceflight, NASA's Mars Climate Orbiter came in too low as it approached Mars in 1999, burning up in the atmosphere. The problem was traced to the fact that two groups of engineers had used different sets of units when computing the Mars orbit insertion maneuver: one engineering group had been using imperial units (feet and pounds), while the other group had been using metric (meters and Newtons).

Imagine proposing and building a spacecraft and flying a mission successfully to another planet, only to die so close to its goal? These are the nightmares that cause spacecraft operators to wake up sweating in the middle of the night and to obsessively check and recheck every aspect of every plan for their birds. It's sobering to realize, but even with a team of really smart, dedicated people, things sometimes still go fatally wrong. Fortunately for New Horizons, and the future world of 2015 that would fall in love with Pluto when New Horizons revealed it, Alice Bowman's team operated their spacecraft flawlessly during its checkout.

At first after launch, time had slowed down as her flight team got used to flying the spacecraft. But as they gained confidence and more and more systems checked out, and New Horizons continued to perform well, time sped back up.

But not everything went perfectly. Barely weeks from Earth, the spacecraft-guidance-system engineer, Gabe Rogers, noticed that one pair of thrusters was firing many times more frequently than it should. The problem was traced to a design error made years before, on a single spreadsheet, that caused these thrusters to be purchased with the wrong specifications. An engineer had miscalculated the spacecraft's mass and balance properties in that errant spreadsheet, and, unlike thousands of other calculations, this one somehow escaped all the independent engineering reviews. So now, in flight, the thrusters were having to work overtime to compensate for being underspecced, and the thrusters were firing too often. Their design life was rated for about 500,000 uses before failure, and computer models had predicted they would be used no more than about half that by the time they'd explored Pluto. But now, with this hyperactive firing, these thrusters were predicted to be used over a million times before the Pluto flyby was completed. Uh-oh.

To remedy this situation, the team put in place some careful new controls on how often the spacecraft would turn in the future, hoarding the use of the two underspecced thrusters, and also began to "load-share" these thrusters by using their backups. The backups suffered the same problem, but by alternating which sets were used, and by limiting their use for some kinds of maneuvers, the flight control team could husband them to keep both the prime and backup sets well under their 500,000-cycle limit all the way to Pluto. Bullet dodged, lesson learned. But it meant a new wrinkle that the flight team would have to live with for the entire journey.

Another key task in those first weeks of flight was perfecting the trajectory toward the Jupiter-gravity-assist aim point that would propel the spacecraft onward to Pluto. After two weeks of careful radio tracking of New Horizons followed by orbit calculations, it became even more clear that the Atlas had put them on a near-perfect trajectory, and that only a tiny spacecraft engine burn correction would be needed to make it perfect. As a result, trimming the trajectory up after launch would take much less fuel than had been allotted when the propulsion system was designed. Those fuel savings gave them a bonus to put in their pocket, either for an asteroid flyby in June, or for more Kuiper Belt exploration after Pluto. Alan chose the bird in the bush—saving the fuel for the Kuiper Belt, a mission objective—which an asteroid flyby, though soon and seductive, was not.

To fine-tune the trajectory to Jupiter, the navigation team estimated the need to burn the onboard engines to change the spacecraft's speed by just eighteen meters per second, or about 40 miles per hour. Not bad, considering it was traveling almost 40,000 miles per hour! But if this correction wasn't made, that 40 miles per hour would mount up—over all the hours until it got to Jupiter on March 1, 2007—to a whopping 400,000-mile error, which in turn would cause them to miss Pluto by several hundred million miles!

To be surefooted with their first trajectory correction of the mission, the team made the adjustment in two parts. First the spacecraft would make a very small engine burn of about five meters per second, just to suss the engines out and see how everything worked. Then once the telemetry from that burn was downlinked and it was confirmed by engineers that it had gone well, the rest of the maneuver would be completed a few days later, putting New Horizons perfectly on track for the Jupiter aim point to Pluto.

With the spacecraft fully checked out, and the engine maneuvers to get precisely on course completed, it was finally time to check out and test the seven scientific instruments that would be humankind's eyes and ears and sense of smell at Pluto. Each instrument team followed a cautious series of steps to carefully turn on their sensor, make sure its computer worked, that it was operating at the right temperature, and that it could be powered and could communicate with the spacecraft's prime and backup systems. Next, the protective launch doors in front of the three telescopic instruments—Alice, Ralph, and LORRI—were opened. Then each instrument was tested to determine if it was performing as it had back on Earth, pointing them at calibration targets, like specific stars in the sky with known brightnesses.

Over a period of many weeks, all seven scientific instruments checked out perfectly, but a near miss with disaster occurred during the LORRI instrument checkouts. Recall that LORRI is a high-powered telescopic camera. The problem: during one checkout it was accidentally pointed directly at the Sun for a few seconds. Just as you can blind yourself by looking at the Sun through a telescope, LORRI's camera can be blinded in the same way. Of course, there was flight software to prevent this from ever happening, but there was a subtle mistake in the way that software was designed. The way the instrument protection software was written, it said, "Anytime we are pointing at a target, make sure that it's at least 20 degrees from the Sun, or don't point there." What it didn't do was ask whether, at any time during the maneuver to swing to a target, LORRI would sweep across the Sun. Unfortunately, as the team was testing LORRI's performance, the spacecraft did just that, pointing LORRI briefly at the Sun during a turn. LORRI, sensing the Sun's extreme brightness, shut itself off to try to protect itself, but not before the Sun briefly poured right down the barrel of LORRI's telescope.

Thankfully, LORRI escaped damage, but the fact that its protecting against scans across the Sun had been overlooked in the "watchdog" protection software produced a real scare, so the flight control team wrote new software protections for LORRI, and the other instruments as well, to make sure none of them could do anything similar again.

During a lessons-learned review of this incident, the New Horizons Pluto flyby mission manager, Mark Holdridge, said to Alan, "Well, we sure dodged a bullet there. I'm glad we learned this, and knock on wood nothing like it ever bites us again." When Mark said that, Alan looked around mission control and realized that everything there was made of something artificial. There was no wood to knock on! So he bought some kitchen cutting boards and sawed them into a dozen little three-by-three-inch blocks. Then he placed a New Horizons mission sticker on the top of each one and attached a little plaque on the bottom of each that read: "Anytime you need to knock on wood for New Horizons, here it is! Keep this until 2015!" Alan then sent a dozen of these little wooden blocks to New Horizons mission control, where they sat on the consoles and office desks throughout the entire flight to Pluto.

Now months into flight, things seemed to be going pretty well. But something odd was brewing in Europe that the New Horizons team did not anticipate.

THE ASTRONOMERS EJECT PLUTO—2006

In August of 2006, barely seven months after the launch of New Horizons, a meeting of an astronomer's organization called the International Astronomical Union (or IAU) was held in Prague. At that meeting the definition of the word "planet" came up for a series of votes.

The Kuiper Belt was turning out to be full of small icy planets that were much like Pluto and were neither gas giants like Jupiter, ice giants like Neptune, or rocky "terrestrial planets" like Venus, Earth, and Mars. Pluto, it turned out, was not alone out there as a large body; it was simply the first discovered and brightest (and therefore the easiest to detect) of this new class of worlds. At the same time, new planets of many other types were being discovered around faraway stars. Most of these were very large planets—as large as or larger than Jupiter. The discovery of smaller planets around other stars was limited by technology, but it was (and still is) widely expected that smaller planets would also be discovered around distant stars in coming years.

Many planetary scientists had long been referring to the rich harvest of newly discovered small planets in the Kuiper Belt as "dwarf planets," a term Alan coined in a 1991 research paper mathematically calculating that the solar system might contain as many as one thousand of them. He chose the term "dwarf planet" in analogy to the well accepted astronomical term "dwarf stars," like the Sun, that are the most common type of stars in the universe.

Then, in 2005, the discovery was reported of a Kuiper Belt planet later named Eris that was thought by its discoverer, Caltech scientist Mike Brown, to be slightly larger than Pluto (later this turned out to be wrong). This in turn resulted in the IAU appointing a planet-definition committee, which included the award-winning science writer Dava Sobel, along with six eminent astronomers. After long deliberations and debates, this august committee proposed a simple and straightforward planet definition: a planet is an object in orbit around any star that is large enough for gravity to make round but not so massive that it ignites in nuclear fusion to become a star. In this scheme, dwarf planets in the Kuiper Belt were recognized as a new class of small planets in line with what many planetary scientists thought.

But what happened next was more than just a little strange. Back in

1980 a British astronomer named Brian Marsden had famously told Clyde Tombaugh that Pluto wasn't a planet in his view, and that it would be his mission to erase Tombaugh's legacy by having Pluto reclassified as an asteroid. No one we've asked remembers why Marsden felt so strongly about this, but people did report that, for some reason, Marsden didn't like Tombaugh. Then, at the 2006 IAU meeting, a gaggle of astronomers led by Marsden procedurally objected to the IAU committee's newly proposed planet definition. Next followed a series of hastily drawn up amendments and redefinitions, all of which were voted down. But on the final day of the weeklong meeting, when most of the attendees had already left the meeting (only 4 percent of the IAU membership remained), those tired few still in Prague voted on a newly proposed definition over the carefully considered one drafted by their own planet definition committee.

Unfortunately, the definition that they voted on was sloppy, awkward, and inelegant, and resulted in Pluto and all dwarf planets, along with all the planets around other stars, being cast out. The hastily arranged voting process the IAU used that day has since nearly universally been regarded as flawed, and the definition it adopted is disliked both by many astronomers, and even more so, by a wide cross section of planetary experts—planetary scientists.

The adopted IAU definition is flawed on multiple grounds. For example, it contains the stipulation that a planet must orbit *our* Sun. This is silly, as it ignores the marvelous discovery that our universe is full of exoplanets, circling nearly every star. Thus, the IAU vote defined "planet" in a way that excludes virtually all the planets in the universe. The IAU further defined "planet" in such a way as to deliberately manage the number of planets in our solar system to be small, on the rationale that schoolchildren would not be saddled with memorizing long lists of planet names if planets were plentiful. (Yes, this argument was actually made with straight faces!) That was accomplished by demanding that a planet must have "cleared its zone" in the solar system. This is a strange way of thinking. If one wants to determine if something is a planet, one should consider its attributes rather than what other bodies it is near or not near in space. But the IAU definition doesn't consider what the object looks like or what its main properties are: for example, whether it has an atmosphere or moons or mountains or oceans. What's key in their definition is where the object is located and what is orbiting or not orbiting near it. By this definition, if the Earth itself was surrounded by a swarm of debris (which it was for its first 500 million years of existence, and arguably still is, even today), it would not be considered a planet.

Adding direct insult to their flawed definition, a final stipulation added by Marsden's group at the end of the IAU's resolution was the vindictive and linguistically nonsensical statement "A dwarf planet is not a planet." With that, Marsden had accomplished his longstanding goal: Pluto would no longer be a planet in the eyes of astronomers, or in astronomy texts, and Clyde Tombaugh's pioneering legacy would essentially be erased.

The IAU vote created a firestorm in the press, primarily focusing on Pluto's "demotion," which is not a neutral term; as "demotion" implies a diminishment of status, a lessening of importance.

It quickly became clear that there had been an effort to cast out the new crop of planets, to remove Pluto and its counterparts from the pantheon of important bodies in the solar system.

When word of the astronomers' vote in Prague reached the New Horizons team, reactions ranged from indifferent ("Who cares what astronomers think? They're not the experts in this."), to bemused, to annoyed, to seriously pissed off. As Fran Bagenal succinctly put it, "Dwarf people are people. Dwarf planets are planets. End of argument."

Many planetary scientists found it particularly annoying that mainstream press outlets seemed to report the reclassification as a fait accompli, accepting without question the authority of the IAU, an organization composed mostly of astronomers, rather than planetary scientists, to define a commonly used word like planet.

Within two weeks of the astronomers' vote, hundreds of planetary scientists—more than all the astronomers who had voted in Prague—signed a petition stating that the IAU's definition was so flawed they would not use it. The press largely also ignored this, and we do not understand why. But as a result, many in the public began to picture Pluto more or less as an asteroid, rather than the small planet it is.

LONG JOURNEY AHEAD

The astronomers' folly aside, the New Horizons team had a great deal of work ahead of them in 2006. In total, their nearly decade-long cruise to Pluto was divided into two distinct cruise phases, each with its own character, work, and pace. The 13-month sprint to Jupiter was jam-packed with spacecraft commissioning, early course corrections, instrument commissioning and calibrations, and the flurry of activity to plan the Jupiter flyby. After Jupiter would be an eight-year-long flight to Pluto, in which the spacecraft would hibernate most of each year while the mission team would plan the flyby of Pluto. Years before, when Tom Coughlin retired as New Horizons project manager and Glen Fountain took over his position, Alan named these two flight phases in honor of the mission's two project managers: the flight to Jupiter was dubbed "Tom's Cruise"; the flight onward to Pluto became "Glen's Glide."

With New Horizons on its long voyage across space, the project team shrank dramatically. In the four-year buildup to launch, more than twenty-five hundred people were involved in building, testing, and launching the spacecraft, its ground system, its RTG, and its rocket. But within a month after launch, the majority of the staff were no longer needed, and went to work on other projects. The big city that was New Horizons was reduced to a small town.

During the long years of flight to Pluto, only a skeleton crew of flight controllers and planners, a handful of engineering "systems leads," the two dozen members of the science team, their instrument engineering staffs, and a small management gaggle was needed. Alan recalls, "Just weeks after launch nearly everyone went their own way, and the project was reduced to a little crowd of about fifty belly buttons. All of a sudden I looked around and it hit me: there are just a few of us—a tiny team— and we're the entire crew that's going to fly this thing for a decade and 3 billion miles and plan the flyby of a new planet."

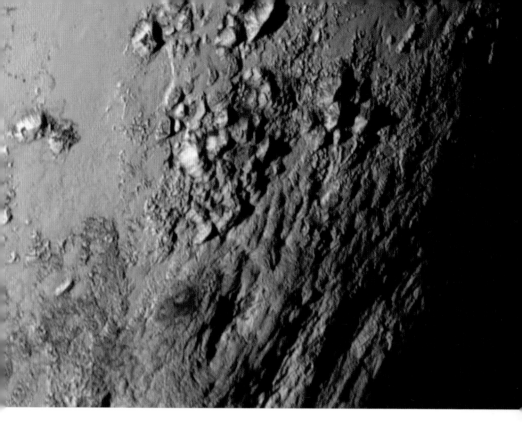

ABOVE: What team members were looking at on John Spencer's computer: the first hi-resolution image shows water ice mountains and a possible ice volcano on Pluto. (NASA)

BELOW: Members of the rock band Styx at APL during the flyby with Mark Showalter, discoverer of Pluto's moon Styx. (NASA/Joel Kowsky)

ABOVE: Scientists and NASA spokesperson Dwayne Brown (in front) celebrate at the moment when the crucial "phone home" signal was received from the spacecraft. (Henry Throop)

BELOW: Alice Bowman and Alan Stern embrace moments after the "phone home" signal from New Horizons revealed that their spacecraft had survived and succeeded at Pluto. (NASA TV)

ABOVE: Queen guitarist and astrophysicist Brian May with team members at the Pluto encounter. (© Michael Soluri/michaelsoluri.com)

AT LEFT: Queen guitarist and astrophysicist Brian May and science team member Leslie Young look at a brand-new stereogram of Pluto, which May put together at the flyby. (Henry Throop)

BELOW: Senator Barbara Mikulski, whose support was pivotal in making New Horizons a reality, speaks at APL during the encounter. (NASA/Bill Ingalls)

ABOVE: The victorious New Horizons core mission
team gather at APL two weeks after flyby.
(© Michael Soluri/michaelsoluri.com)

BELOW: Press conference at APL on July 15, 2017.
From left: Dwayne Brown (NASA), Alan Stern, Hal
Weaver, Will Grundy, Cathy Olkin, John Spencer.
(NASA/Bill Ingalls)

The special issue of *Science* magazine, with first results from the New Horizons flyby of the Pluto system, published just ninety-four days after flyby. (Reprinted with permission from AAAS)

Science

$10
16 OCTOBER 2015
sciencemag.org

AAAS

Flying past Pluto
New Horizons finds surprises at Pluto and Charon *pp. 260 & 292*

ABOVE (true color): The "fail safe" image of Pluto, the last and most detailed image taken on July 13, 2015, the day before closest approach. (NASA)

OPPOSITE TOP: Fifteen minutes after closest approach, New Horizons shot this dramatic image showing a complex surface with icy mountains and smooth plains, and layers of haze over a crescent Pluto. (NASA)

OPPOSITE BOTTOM: This image shows the wide range of surface ages and terrain types on Pluto, from dark, ancient heavily cratered terrain at the bottom to bright, young, crater fields of nitrogen ice at the top. (NASA)

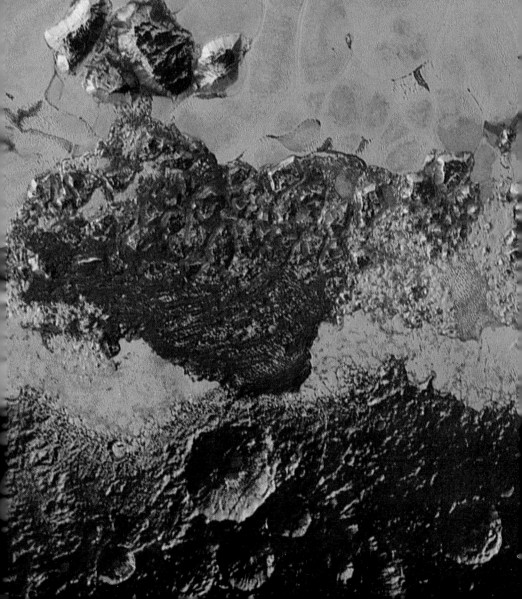

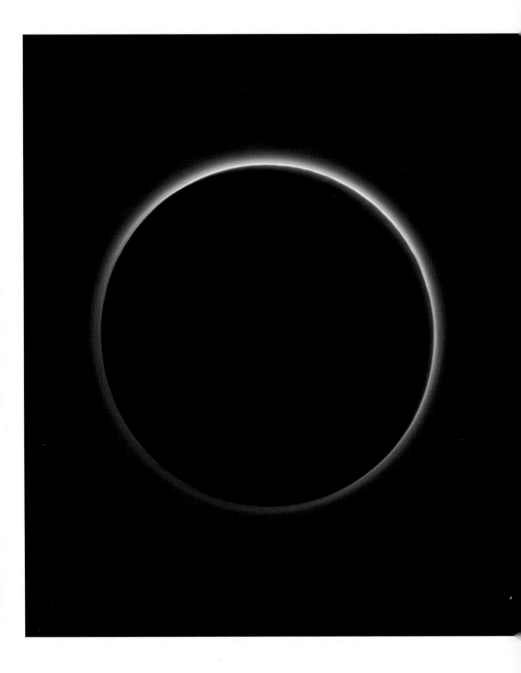

A look back toward Pluto as New Horizons
receded outward from the Sun revealed a
stunning ring of blue sky surrounding the
planet, just like Earth. (NASA)

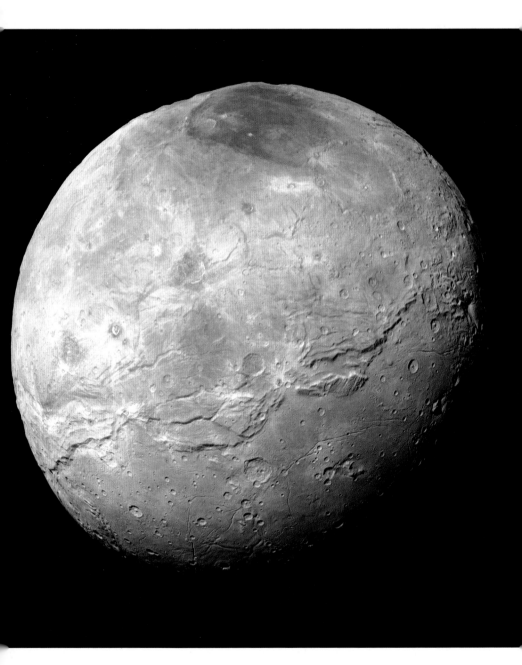

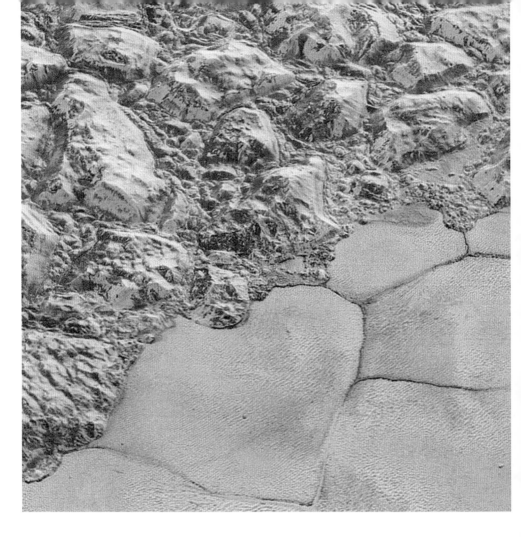

Voyager
Terra

Hayabusa Terra

Djanggawul
Fossae

Al-Idrisi
Montes

Burney

Sputnik
Planitia

Sleipnir
Fossae

Tartarus

Dorsa

Elliot

Hillary
Montes

Virgil Fossae

Tenzing
Montes

Adlivun
Cavus

**Tombaugh
Regio**

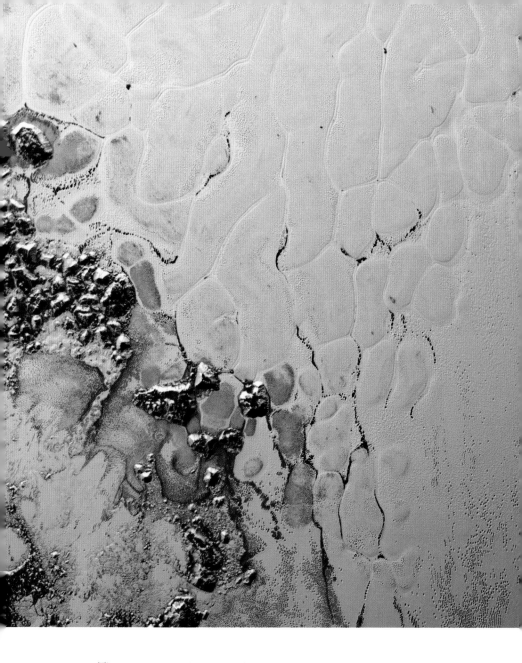

OPPOSITE TOP: The mountainous shoreline of Sputnik Planitia shows peaks of water ice towering above flat plains of nitrogen ice with a surface texture that may indicate dune fields created in epochs of higher atmospheric pressure on Pluto. (NASA)

OPPOSITE BOTTOM (color enhanced): The first set of feature names on Pluto proposed by the New Horizons Team and formally adopted. (NASA)

THIS PAGE (color enhanced): The young surface of Sputnik Planum (the left side of the heart) shows a geological pattern of cells caused by the churning motion of convection in the nitrogen ice. (NASA)

ABOVE (color enhanced): Mountains capped with bright methane snow in the Cthulhu region of Pluto. (NASA/Alex Parker)

BELOW (color enhanced): This enhanced color image of the area near Pluto's north pole shows a region cut by deep canyons and valleys. (NASA)

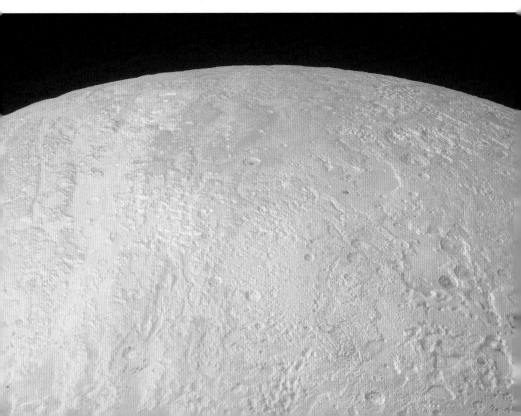

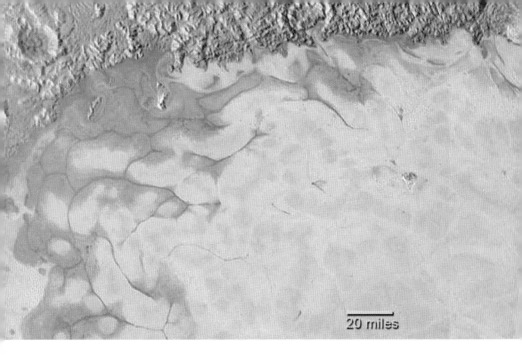

ABOVE: Patterns on the edge of Sputnik Planum closely resemble large glaciers on Earth, indicating flowing nitrogen ice. (NASA)

BELOW: The four small moons of Pluto, shown to scale with the crescent (below) of Pluto's giant moon Charon. (NASA)

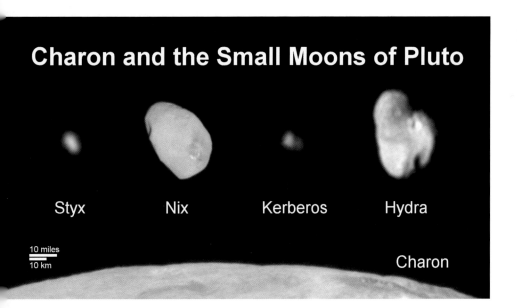

TOP (color enhanced): A detailed New Horizons view of Sputnik Planum shows a wide variety of terrains, including smooth plains surrounded by sharp mountains, deep canyons, and dark, cratered areas. (NASA)

BELOW: Single frame of a double planet. The highest resolution image from New Horizons, which captures both Charon (left) and Pluto (right) in enhanced color. Here the two bodies are realistically shown to scale with one another and also to scale with their true separation distance. (NASA)

AT LEFT: This feature, about 30 kilometers (20 miles) long, is suspected to be a frozen lake of liquid nitrogen, hinting at a time in the past when surface pressure may have been much higher. (NASA)

BELOW (color enhanced): This high resolution image shows the Southeastern edge of Sputnik Planitia where it meets the very rugged darker terrain of Krun Macula (the westernmost of the "brass knuckles"), which rises a mile and a half above the plains. The bright surface of Sputnik is dotted with kilometer-scale pits probably caused by sublimation of nitrogen ice. (NASA/ John Spencer, Paul Schenk, Pam Engebretson)

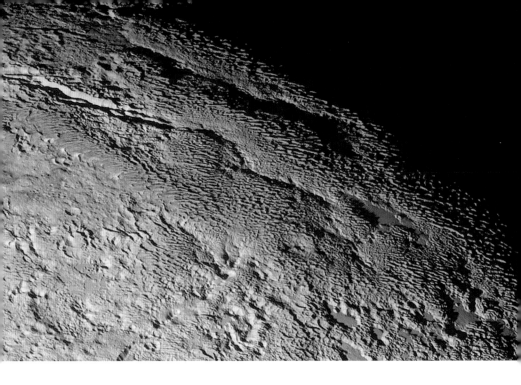

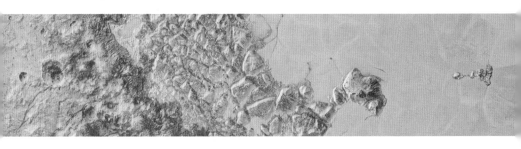

TOP (color enhanced): This strange "bladed terrain" stretching hundreds of miles on Pluto is seen nowhere else in the solar system. It is composed of jagged spires of methane ice an average of twelve hundred feet high. (NASA)

ABOVE (color enhanced): Part of a high-resolution color mosaic of Pluto taken just before closest approach shows features as small as 270 yards across, from craters to faulted mountain blocks to the textured surface of Sputnik Planum. (NASA)

You might think given the vast distance between Earth and Pluto, and all the years of flight time ahead, that boredom or impatience could have been a problem for the mission team. But in actuality, advances in automation and the plan to hibernate the spacecraft for most of the journey meant the New Horizons mission team was almost ten times smaller than the vast Voyager crew of 450. As a result, the stripped-down New Horizons team remained remarkably busy for ten straight years.

This was particularly true during the first leg of the journey. The thirteen-month trip out to Jupiter was lightning fast, and there was so much to do. Hal Weaver remembers:

> There was no rest for the weary. After we got the spacecraft checked out, the first course corrections made, and the instrument payload checked out, the schedule was still hairy, because we had to plan and execute a complex Jupiter flyby almost immediately. We had to navigate to the right keyhole in space near Jupiter to fly on to Pluto, and we also wanted to have a practice at Jupiter for all of the steps and through the process that we would be using for planning the Pluto encounter, and we wanted to plan a gangbuster science flyby at Jupiter as well. All of that had to be prepared and tested and executed within just thirteen months after launch.

A longer-term concern was retaining the corporate memory over the nearly ten years of flight ahead—all the fine details about how the spacecraft and instruments are constructed and operated. This was especially important because more than nine out of every ten people originally on the project went on to other projects after New Horizons launched. What would happen a decade hence when the mission had to be staffed up again for the flyby, with many new people who had never been a part of designing and building the bird? To counter this and other prudent concerns about the long road ahead, team members documented as much as possible about every aspect of the spacecraft and its mission control. They also made training plans for those they would hire eight or nine years in the future, and they created spare parts inventories for the mission control and spacecraft simulators. They even created videos describing (classroom style) how every detail of the spacecraft and mission control worked.

This work to make sure all of that corporate knowledge was captured while still fresh, and made ready for that distant year of 2015, was in addition to the work to learn to fly New Horizons, to get it and its scientific instruments checked out, and to plan the Jupiter encounter. Hal Weaver was right: there was no rest for the weary flight team during the entirety of 2006 and most of 2007.

BY JUPITER

For the science and mission operations teams, the largest single activity of late 2006 was planning their 2007 Jupiter flyby. There were three separate reasons why this flyby was key. First and foremost, they had to fly through a tiny window in space near the precise aim point at Jupiter, and at just the right moment, to get the gravity assist needed to target Pluto. If that didn't succeed, the spacecraft would be headed elsewhere, and it would be game over for Pluto exploration. Second, Jupiter was the only encounter with anything before Pluto, so that flyby was the one opportunity to practice an actual planetary encounter, running all the spacecraft systems through their flyby paces, and trying out all the instruments on a planet and its moons. Among the many examples of this were tests of the optical navigation capabilities that would be needed at Pluto. The Jupiter flyby would be used to perfect how to determine the precise range and aim point offset to their flyby target by comparing the changing position of Jupiter against the faraway stars in successive images.

Alan told the team about the practice they would get on the Jupiter flyby: "Our objective here is to learn as much as we can about the spacecraft by putting it through its paces. At Pluto our objective will be to learn about Pluto, and I don't want to learn a thing about the spacecraft there—we need to learn all those lessons here, at Jupiter, so that the Pluto flyby is flawless." The final Jupiter priority: to take advantage of the flyby—the first of Jupiter since NASA's Cassini flew by in 2000 on its way to Saturn—to learn more about that planet, its largest moons, and its gigantic magnetospheric cocoon of charged radiation.

New Horizons was in many ways more capable than any previous spacecraft sent to Jupiter, and the New Horizons team was psyched to see what new things they could learn at Jupiter using their seven state-of-the-art scientific instruments. There had been previous close looks and even long-term observations of the Jupiter system, but it's a complex and ever-changing place, so it was valuable to have another exploration of it in 2007, particularly one bringing powerful new instrumentation to bear.

Planetary geologist Jeff Moore, the New Horizons Geology and Geophysics team lead, honchoed the New Horizons Jupiter Encounter Science Team (called JEST). Jeff was a veteran at Jupiter exploration, having served as a graduate student on Voyager, and then on NASA's Galileo Jupiter orbiter in the 1990s.

Planning for the New Horizons Jupiter encounter went into high gear in the fall of 2006, just after the instrument checkouts and calibrations were completed. Because their Jupiter-to-Pluto aim point was actually very far—4 million miles—from Jupiter, New Horizons wouldn't be able to swoop close to any of Jupiter's moons, which all orbited much closer in. Nonetheless, many important observations of them could be made using the onboard telescopic instruments.

One goal was to create a multi-week record of volcanic activity on Jupiter's planet-size moon Io, the most volcanically active world in the solar system (outperforming even Earth). Another was to create a similar, multi-week record of Jupiter's relentless storm systems. Neither of these objectives had been accomplished by the Galileo mission because of problems it had returning data due to its crippled spacecraft antenna.

And beyond what the telescopic instruments could contribute, in a stroke of luck, the post-Jupiter path that New Horizons would take on its way to Pluto would fly literally 100 million miles down the tail of Jupiter's roiling magnetosphere. This would make possible a whole

new and pioneering class of observations of the giant planet's magnetosphere with SWAP and PEPSSI. Nothing like that had ever before
been accomplished, because no other mission had flown so deeply down
any giant planet's magnetosphere. In all, almost 700 observations of the
Jupiter system were planned, involving all seven instruments on New
Horizons.

The Jupiter flyby, which entailed observations while approaching,
at, and then departing Jupiter, spanned January to June of 2007, and
was a spectacular success. The navigation team flew through the Pluto
aim point with precision. Additionally, all of the many spacecraft and
instrument tests went well, and New Horizons collected a wide range
of scientific observations that ultimately landed it on the cover of the
prestigious research journal *Science*.

Perhaps the most enchanting result at Jupiter came about through
simple, dumb luck. During the flyby planning, the team realized that
it would be able to make some key new studies of Jupiter's thin, dusty
rings. However, no one on the New Horizons team was an expert in
this scientific area. So planetary rings specialist Mark Showalter was
brought on board to help. Showalter designed a five-frame movie to
be taken by the LORRI camera at just the time when Jupiter's moon
Io would be crossing in front of the main Jovian ring. His idea was to
make high-resolution maps of the ring as Io occulted it. But when
these images came down to Earth, the team was astounded to find
that in addition to accomplishing that objective, the image sequence
had caught one of Io's large volcanoes, Tvashtar, in the act of erupt-
ing, spouting a giant plume off the north pole of Io. Volcanoes on Io
had been photographed many times before, and both the Voyagers
and Galileo had caught dramatic shots of volcanic plumes shooting
off Io into space! But before New Horizons, no time-lapse sequence
of any volcano off Earth had ever been captured. The result was remarkable—a shimmering fountain of material being ejected from Io
high into space and falling ballistically back down toward its surface.
It was both scientifically valuable and mesmerizing.

JUPITER'S STING

We pointed out earlier how radiation can negatively affect the electronics on board a spacecraft. Just days after closest approach to Jupiter, there was an unexpected and unwelcome event, which had to do with just that. New Horizons was stricken with a "safing event," caused when something goes wrong on board the spacecraft and triggers it to go into "safe mode"—a configuration in which the errant system is turned off, its backup system is engaged, and the spacecraft radios Earth that something is wrong and it is awaiting instructions. Safe mode is common to many spacecraft; it is designed to keep spacecraft from doing anything that will cause more harm or exacerbate a problem, until the situation can be diagnosed—and resolved—from the ground.

When Alice Bowman's mission ops team got telemetry back from New Horizons revealing this problem, it was found to be due to the main flight computer, which had rebooted itself. Something very similar had occurred on approach to Jupiter. A second occurrence in so short a period, just three months, was worrisome. Then, several months later, the same thing happened again—another unexplained main computer reboot, and another safe-mode entry. Initially some on the team were concerned that the main computer was failing and might not make it to Pluto. As more such events occurred later in the mission, they were found to be spaced at longer and longer intervals. The engineering team developed a theory that the main computer was not failing, that instead the reset events had been caused by radiation damage associated with Jupiter's powerful magnetosphere of charged particle radiation. They hypothesized that over time the circuits causing the resets were healing from the radiation damage. By the time New Horizons reached Pluto, no computer reset had occurred in years: Jupiter's sting had worn off.

TO HIBERNATE

If a person bought a television in January of 2006 and was expecting it to be operating in mid-2015, would it be more likely to meet their

2015 needs if they left it on continuously for those nine and a half years, or if it was turned off and only checked out occasionally until 2015 came? From an engineering perspective, the latter course is more likely to succeed, and this is the concept behind hibernating the spacecraft, the next big task for the New Horizons team after Jupiter.

Hibernation is one of the pioneering and truly innovative aspects of New Horizons. The idea behind it is to save wear and tear on most spacecraft systems by turning them off for most of each year en route from Jupiter to Pluto. Thanks to hibernation, by the time New Horizons got to Pluto it would be 9.5 years old on the clock, but most of its primary systems would only have about 3.5 years of "on time" clocked against them. Those spacecraft systems would, in effect, be many years "younger" than their actual age, less worn out, and more likely to be operating at peak performance at the Pluto flyby.

The software designed to hibernate New Horizons was written before launch but it had been tested for only a few days when the spacecraft was at Goddard. Because the flight hibernation periods would typically last for months on end, the plan was to ease into it carefully, to make sure the spacecraft didn't get into any trouble while being left in this hibernating state for longer and longer periods.

The first test was for only one week in the summer of 2007. When the craft came out of hibernation it sent all the resulting, stored telemetry down to Earth for engineers to evaluate how it went, and everything looked good. So next the spacecraft was commanded to hibernate for a few weeks. Then, after the spacecraft engineers looked at that data and saw that longer test also had gone well, they tried hibernating for ten weeks, then for four months. As the team became more and more confident, they increased the hibernations to as much as seven months at a time.

Each time the spacecraft hibernated, from 2007 to 2014, the mission team could turn from babysitting their bird to focus on the massive task of Pluto flyby planning, which we will describe in the next chapter. Between hibernation periods, the spacecraft would be woken up to be thoroughly checked out, to conduct instrument calibrations and en route science observations, and, from time to time, receive software upgrades to correct bugs or add new capabilities. One key up-grade allowed the SDC, SWAP, and PEPSSI instruments to remain on, taking data while in hibernation so they could trace out the solar system's dust and charged-particle environment all the way to the Kuiper Belt, a scientific bonanza en route to Pluto.

11

BATTLE PLAN PLUTO

PLANNING THE INVASION

Early in 2008, project manager Glen Fountain laid out a Pluto flyby planning schedule that would unfold over six years and involve virtually every person on the New Horizons flight team. Most of the flyby planning was slated to take place during lulls in mission operations, when the spacecraft was in hibernation and needed little tending.

The undertaking was massive for such a small team. First, a series of technical design studies of the flyby's specifics were needed for the science and mission operations teams to be ready to work on detailed objectives for the flyby. All of that was needed before the comprehensive design of hundreds of flyby observations of Pluto and its moons by the seven scientific instruments aboard New Horizons could be made. This in turn would have to be followed by exhaustive testing of all flyby operations, and an enormous effort to generate procedures to handle hundreds of possible malfunctions the spacecraft and its ground team had to be ready for. That in turn was then followed by flyby simulations and rigorous project and NASA technical reviews to inspect, critique, and insert recommendations to improve every aspect of the flyby plan.

It's interesting to compare the high-tech, computer-drenched twenty-first-century flyby planning effort by New Horizons, with the comparatively stone age mainframe computer–era planning for Voyager flybys back in the 1970s and 1980s. The Voyager team had just a few years to plan each giant-planet flyby, but their team was huge. Each Voyager flyby involved almost five hundred people and took most of three years to plan. New Horizons would take six years to plan its Pluto flyby, but would involve a crew of just fifty people—barely 10 percent the staffing level of Voyager.

The Pluto flyby planning czar was Leslie Young, who had undertaken a skeletal version of Pluto flyby planning as a part of the New Horizons proposal to NASA in 2001. Having seen her impressive flyby planning work, drive, and attention to detail on the proposal, Alan asked her in 2008 to lead the planning of the real thing.

When New Horizons was proposed, Alan had organized the science team into crosscutting "discipline" theme teams, focusing on different aspects of Pluto system science. This prevented the usual balkanization into instrument teams that had caused many other planetary missions to develop warring instrument camps. The four New Horizons theme teams were: the Geology and Geophysics team (applying the techniques and insights of the geology of Earth and other planets toward understanding the structures and motions of Pluto's surface, and interior, as well as its satellites), led by theme team leader Jeff Moore of NASA's Ames Research Center; the Composition team (determining what the surfaces of Pluto and its satellites are made of), with theme team leader Will Grundy of Lowell Observatory; the Atmospheres team (focusing on measurements of Pluto's atmosphere), with theme team leader Randy Gladstone of SwRI; and the Plasma and Particles team (understanding the way that Pluto and its moons interact with the solar wind and what the ionized gases escaping from Pluto's atmosphere are made of), which was led by Fran Bagenal of the University of Colorado. All four groups reported to Leslie for flyby planning and comprised what was dubbed the overall Pluto Encounter Planning (or PEP) team (sometimes jokingly called the "PEP squad").

Several other New Horizons teams were also key to the flyby planning: mission operations planning was led by Alice Bowman; mission design and navigation was led by Mark Holdridge; and spacecraft engineering was led by Chris Hersman. A big part of the spacecraft engineering team's work involved making sure that no onboard resource, like fuel or power or data storage, ever exceeded its safe limit. As project manager, Glen Fountain herded all these cats to stay on schedule and on budget. Alan wore a few different hats during the encounter planning. He led the Alice and Ralph instruments as their instrument PI; he also sat on Leslie's PEP executive board (which also included Cathy Olkin and John Spencer); and as mission PI he was the ultimate reviewer, critiquer, and approver of all aspects of flyby planning, contingency planning, and team training.

FINDING JUST THE RIGHT DISTANCE AND TIME FOR A FLYBY

Before the New Horizons team could put any kind of detailed flyby plan together, there were two big decisions to make: exactly when the spacecraft should fly through the Pluto system and at exactly what distance. So beginning early in 2008, Leslie captained detailed studies to pick the best day and best altitude to fly by Pluto.

There was enough fuel aboard New Horizons to change the date of the flyby by up to a few weeks from its nominal date in mid-July of 2015, and Alan wanted to get the most out of the flyby by finding the absolute most optimal date. So Leslie and the PEP team evaluated every possible factor, from which parts of Pluto the spacecraft would fly over on each given day (as Pluto rotates on its axis once every 6.4 Earth days), to the distances of every moon from New Horizons on each possible flyby day, to which radio tracking stations would be in position on Earth to perform the radio science experiments to measure Pluto's atmospheric pressure and radar reflectivity. In all, Leslie's arrival date study looked at more than a dozen factors that varied across each day in late June and all of July in 2015 to choose the best possible arrival day. No trajectory would be perfect for every factor, so deciding among them involved a complex series of trade-offs. In the end, Leslie's team recommended, and Alan approved, July 14, in part because on that date the spacecraft would fly over Pluto's brightest terrain spot (which also was known to have an unusual composition). But July 14 also gave the best combination of satellite science, and it cost the least fuel—because that was the date they had originally targeted at Jupiter flyby, which was a bonus because Alan also wanted to save fuel for flybys of bodies farther out in the Kuiper Belt after Pluto.

After settling on July 14 as the encounter date, Leslie's team then began looking at flyby closest-approach distances. The most important science observations would be of Pluto itself, so the closest-approach distance to Pluto was key; but the distance to each satellite was also important. The four science discipline theme teams began by weighing how well each of their detailed Pluto scientific objectives would be accomplished for a range of possible Pluto closest-approach distances from about three thousand to twenty thousand kilometers, with corresponding ranges for each satellite of about twenty-eight thousand to almost eighty thousand kilometers. Going really close helped the plasma instruments detect more phenomena, but created problems for the cameras. (You might think the camera teams would want the closest view possible, but at a flyby velocity of over 30,000 miles per hour, going too close meant that images from too close would be smeared by the spacecraft's blazing speed.) Dozens of factors were considered. In the end, flying by at a distance of 7,800 miles, deep inside the orbits of all of Pluto's moons, was selected as the best overall way to satisfy the competing desires of the four science discipline teams.

Calculations showed that for the 7,800-mile closest approach flyby to work—for all the images to be properly centered on their targets—the spacecraft had to arrive no more than nine minutes off target after its 9.5-year journey, equivalent to a cross-country airline flight from Los Angeles to New York landing within four milliseconds of its planned time! New Horizons also had to reach the Pluto aim point at closest approach no more than about 60 miles off course after the 3-billion-mile journey from Earth; that 60 miles is about the size of metropolitan Washington, DC, and the spacecraft had to hit this target in a journey traveling all the way from Earth to Pluto. This was the equivalent of hitting a golf ball from L.A. to New York and landing it in a target the size of a soup can! That made for a tall order.

FLYBY PLANNING BEGINS

To begin planning the multi-month flyby itself, the encounter was broken into stages. It would begin in January of 2015, still six months and almost 200 million miles from Pluto, with what was dubbed Approach Phase 1, or AP1. AP1 was primarily needed to gather navigation imagery to home in on Pluto, but it also included measurements of the environment that Pluto orbits in using the SWAP and PEPSSI plasma instruments and the Student Dust Counter. This far out, Pluto would just be a dot in the distance. With the start of AP2 (Approach Phase 2) in April of 2015, the distance to Pluto would be cut in half compared to the start of AP1. So in addition to continuing the same kinds of activities as in AP1, the spacecraft would then be able to start to see Pluto as well as Hubble could see it from Earth. After that, images would get progressively better as each week went by—so the first scientifically useful Pluto observations were planned for AP2.

AP3 would begin much closer to Pluto—in mid-June, and it would last only three weeks but would include an intensive imaging campaign of Pluto and its moons as they circled the planet; it would also include the first compositional observations of Pluto and Charon by New Horizons, and intensive searches for new moons and even rings. After AP3, the so-called Core of the encounter would then begin, just seven days before flyby closest approach. Core would last until two days after closest approach. The Core flyby would then be followed by three DP's, or departure phases, continuing until October of 2015.

The three approach phases, the three departure phases, and the Core itself would each be planned separately and with different levels of rigor. The phases closest to Pluto would be prepared farther in advance so that much higher levels of testing could be done, reflecting their greater importance to the overall science return.

Each of the flyby's phases was in turn then broken down into one to several long "command sequences," each consisting of thousands of computer instructions to run the spacecraft, point it at the various targets in the Pluto system, operate the instruments, and store each data set. But before these command sequences could be written, Leslie's PEP team designed over one hundred measurement techniques (MTs) to cover all the scientific objectives of the flyby. Each MT laid out where to point, what instrument and mode to use, at what distance from the target, which data recorder to store the results on, and so forth. Each MT was assigned a "champion"—a relevant expert on the science team who was responsible for designing it. Alan insisted that after each champion designed an MT, the MTs would be scrupulously reviewed by PEP to look for deficiencies and opportunities for improvements.

The Core phase of the Pluto encounter would be when the actual flyby and closest approach to Pluto and all five of its moons would occur. During this nine-day-long beehive of onboard activity, the spacecraft would be placed in a special software mode called "Encounter Mode," which would prevent onboard problems from ruining the encounter by stopping activities and calling home to Earth for help. If spacecraft engineers had a more playful sense of humor about such dire emergencies, Encounter Mode might have well been called the "Do not disturb!" or "Don't bother me, I'm busy!" mode.

In routine cruise flight, when the spacecraft detects an onboard problem, like the computer reboots described in Chapter 10, its autonomy software is designed to triage the problem, fixing the immediate danger (e.g., closing fuel valves if a fuel leak is detected), informing Earth of the problem, and then enter Safe Mode, stopping all further activities until new instructions arrive from the New Horizons Mission Operations Center. This kind of response is designed to keep the spacecraft from getting into any deeper danger until engineers on Earth have time to analyze the problem themselves and devise a complete response. But during the days closest to Pluto, such a process would be countereffective, because if the spacecraft shut down activities until help arrived—which would take a minimum of half a day from faraway Earth—whole swaths of irreplaceable science observations would be lost. So Encounter Mode was designed to handle problems differently when they matter most—at Pluto. In Encounter Mode, the spacecraft still triages its problem, but then it just goes back to the next step on its timeline, continuing to implement the observation plan. The logic of Encounter Mode is that while at Pluto it's better to try to take data, even if the spacecraft has a problem, than to stop and wait for help from Earth. Encounter Mode had been planned ever since the proposal days, but never implemented. As flyby planning began, it was time for Chris Hersman and his spacecraft team to design and build the Encounter Mode software.

At roughly the same time, the detailed planning of the flyby science sequences also began. The almost five hundred observations planned for the flyby involved all seven instruments aboard New Horizons, and addressed almost twenty separate top-level scientific objectives laid out by the New Horizons team and NASA. These objectives included mapping Pluto, imaging all of its satellites, measuring Pluto's atmospheric properties, searching for moons and rings, measuring the temperatures of Pluto and its moons, and much more. Not only would each of the almost 500 observations be individually designed and then built as a sequence of software commands (turn on this instrument, select this mode, point in this direction, store the data, etc.), each would also be thoroughly tested on the New Horizons Operations Simulator (or NHOPS, pronounced "En-Hops"), a high-fidelity electronic replica of the spacecraft and its scientific instruments, housed at APL.

With painstaking care, Leslie and her PEP team designed every single observation using flyby planning tools—software packages—they had built to let them check everything from the expected resolution and signal to noise of any given observation, to whether the instrument pointing was correct, to how forgiving the observation was to the spacecraft being a little off course or Pluto or its moons not being exactly where they had been mathematically predicted to be.

PEP worked most closely with two other teams to do its job: SciOps (science operations), which planned all of the command sequences for the seven scientific instruments, and MOPS (or mission operations), which planned all of the spacecraft's activities, from communications to data storage to interior temperature control to engine burns for course corrections. Together these three groups—PEP, SciOps, and MOPS—carefully choreographed every aspect of the flyby's flight operations.

One might at first think this would be as simple as just laying out where all the observations needed to be placed on the timeline, but the planning required was actually much more complex and nuanced. In effect, the PEP, SciOps, and MOPS teams were playing a kind of twenty-plus-dimensional chess game. For every observation, they had to make sure the spacecraft would never go over power budget, that time was always allocated to maneuver the bird to whatever direction the instruments needed to be pointed, that sufficient onboard data-recorder space was always set aside, and so forth. Literally dozens of factors had to be considered to plan each observation. Additionally, these three teams had to weave into the timeline a backup for each of the most important science observations, to ensure that if the spacecraft or one of the instruments malfunctioned during a critical data take, there was a second chance to get a similar data set.

The teams also devised their plans to be resilient to possible instrument failures. For example, observations were planned to back up Ralph imaging with LORRI imaging, and vice versa. They also planned for each instrument to cycle, from observation to observation, between its prime and its backup electronics, and even to sprinkle instrument reboots throughout the timeline in case any instrument got stuck in a bad mode at any point. All of this had to be scripted down to the finest details, years in advance, because there was simply not time to do all this with a small team during the final year or two before flyby. For the New Horizons team, the brainstorming, planning, review, and testing of the entirety of the six-month flyby consumed most of the years from 2009 to 2012.

If one was planning any analogous set of scientific investigations of Earth, such a detailed and hyper-scrutinized planning might seem neurotically compulsive. But at Pluto, if anything was improperly designed or was not fully thought through, there would not be another chance to get it right. So all this obsessively careful planning, all the checking and counterchecking, was the way to ensure that the mission would actually get the intended goods in humankind's shot at exploring Pluto and its system of moons.

SMASHING BUGS

As leader of the mission, Alan felt as though it was part of his job to look for weaknesses in the flyby plan, ask a lot of questions of his teams, probe their assumptions, and ask for changes to fortify the planning. One of the many weaknesses he spotted and changes he asked for concerned NHOPS.

About the time that the seven various flyby phases were being laid out and architected, Alan became concerned that the NHOPS spacecraft simulator, which was used to test all spacecraft command sequences to weed out bugs, could become a showstopper if it failed in 2015. He just wasn't comfortable with the fact that an NHOPS failure in 2015, when there was little time for a repair, could risk the team's ability to fully test the flyby sequences. A backup, called NHOPS-2 was already in place, but it was a stripped-down version of NHOPS that lacked much of the simulation capability and fidelity of the original. So at Alan's direction, Glen put plans and budget in place to convert NHOPS-2 into a full-up spare to NHOPS, and to test it as thoroughly as had been done with NHOPS, to be sure it would be ready if there ever was a need. Little did he know then, this was a decision that would prove crucial in the final days of the approach to Pluto.

As each of the dozens of command sequences that together comprised the entire flyby were designed and had passed their peer reviews, the MOPS team began running them on the NHOPS spacecraft simulator to see if they would work as expected. Bugs were often found, corrected, and then new NHOPS runs would be scheduled. MOPS repeated this again and again, until every sequence ran completely error free. For the Core load—the crucial nine-day-long chain of sequences that instructed the spacecraft how to execute all actions during the close flyby—that took eight tries. Each of these eight

NHOPS runs of the Core load took the full nine days. Version 1 was called V-1, version 2 was called V-2, and so on. Every time the team found bugs it rewrote the errant parts of the sequence and started the NHOPS run again, from scratch. When the Core finally ran bug free on V-8, the eighth of these nine-day-long NHOPS runs, Alan celebrated by buying a couple cases of little cans of V-8 juice and handed the cans out for each team member to keep as souvenirs of the time-consuming battle to create a completely bug-free Core load.

Once that error-free milestone was achieved, the Core sequence was "locked down" under a rigorous "no change without careful review and approval" process, called "configuration management" (or CM). CM's job was to ensure that an extra level of scrutiny and testing rigor went into any change, no matter how minor. Weekly meetings of a group called the Encounter Change Control Board (or ECCB), were held to evaluate change requests to the Core load and half a dozen other sequences that would run on New Horizons during the period from May to July of 2015. The ECCB was chaired by Alan and staffed by chief engineer Chris Hersman, project manager Glen Fountain, MOPS lead Alice Bowman, senior project scientist Hal Weaver, PEP lead Leslie Young, and encounter manager Mark Holdridge.

PREPARING FOR PROBLEMS

At the same time that all the Pluto flyby command sequences were being developed, the project team also took a look at everything that could possibly go wrong during the encounter and how they or the spacecraft would have to react in order to fix any given situation. This kind of "malfunction procedures" development is common to space missions, and it was crucial for a one-shot opportunity like a Pluto flyby.

The largest effort to prepare for potential problems was led by spacecraft chief engineer Chris Hersman. Hersman, incredibly sharp and meticulous in his attention to detail, and incredibly knowledgeable about all aspects of the bird, made plans for each of 264 potential spacecraft, ground system, and other problems that might arise. This wide-ranging effort, which took three years to flesh out and implement, reached into every aspect of the project. Hersman went far beyond what might go wrong on the spacecraft, looking also at what could go wrong with the team or in mission control itself. So for example, plans were made to train backups for every critical role, in case anyone was unavailable at the flyby due to health problems, car wrecks, or family emergencies. Chris also made sure that detailed checklists were put in place, peer reviewed, and then practiced, for flight controllers to handle dozens of different spacecraft and instrument malfunctions too complex for the onboard autonomy system to handle. Chris even made sure that, in the event that the Mission Operation Center suffered a fire or a terrorist attack (after all, New Horizons would be a high-profile target in 2015), the project had a fully capable, and fully checked-out second mission operations control center to use across the APL campus. Each of Hersman's 264 backup plans was reviewed and critiqued in detail by engineers on the project, and then again by Glen and Alan, over a series of two dozen painstaking, multi-hour meetings that stretched across 2011–2014.

PLANNING THE GROUND GAME

Still another layer to flyby preparations began with the development of a plan for the "ground game" logistics of the flyby, which was going to eventually involve about two hundred people working in close concert over several months. Mark Holdridge, his deputy, Andy Calloway, Glen Fountain, Glen's deputy Peter Bedini, and Alan's assistant, Cindy Conrad, undertook this enormous planning effort. They began by mapping out where each of the almost two hundred people across the country would have to be, and what meetings each would attend, during every single day from January to July of 2015.

This included mapping out every trip to APL by every person involved in flyby operations who did not live in Maryland, and even looking at shift schedules to identify periods when any given individual might not be getting sufficient sleep—owing to the pace of spacecraft activity taking place at odd hours back on Earth. It also included assessing all the office and conference room space needs at APL and reserving all the needed meeting and conference rooms for the 130 or so team members traveling to APL. Cindy Conrad and her assistant, Rayna Tedford, planned who needed what kind of APL access badges and what office supplies the travelers would need while in residence at APL. They even arranged for runners who could go out for meal orders when people didn't have the time to leave APL, and they planned who among the APL staff lived too far away and would have to move into nearby hotels to be available at any time during the few weeks around closest approach.

PRACTICE, PRACTICE

As the arrival at Pluto neared, the project made plans to rehearse and practice as many aspects of the flyby as time, money, and individual schedules could afford.

The centerpiece of this activity was the development of in-flight rehearsals in which New Horizons itself was loaded up with its actual Pluto flyby sequences and instructed to execute them in full. These flyby rehearsal runs took place out there in empty interplanetary space, running the spacecraft through all its paces to make sure that what had worked on the NHOPS spacecraft simulator would also work perfectly on the bird itself during the close flyby.

The first of the spacecraft rehearsals took place in July of 2012 as a two-day "stress test" of the most intensive operations for the period right at closest approach. A second rehearsal was run in July of 2013. That one was a full-up, comprehensive, nine-day test of the entire close-approach Core sequence. This rehearsal even went so far as to involve the entire New Horizons ground control, science, and engineering teams, working the same shifts and performing all the same activities, holding all the same status and decision meetings, and even reporting "progress" to NASA, as they would during the real thing two years hence in 2015. Additionally, all of the remote elements of the flyby team, like the Deep Space Network tracking stations, were involved in this rehearsal. After each of the spacecraft rehearsals, the New Horizons team also undertook a detailed review to search for and repair even the tiniest discrepancies or flaws in what had occurred—on the bird out in space, or down on the ground.

Yet an additional layer of practice for the flyby took the form of ground simulations, called Operational Readiness Tests, or ORTs. The ORTs were elaborate training drills in which various parts of the project undertook multi-hour to multi-day simulations of planned or malfunction-related flyby activities. The navigation team alone held almost a dozen of these as multi-day practice runs, led by Mark Holdridge, each with specific objectives and scoring of how well the team members, their processes, and software tools did. Each "Nav ORT" resulted in a formal action list of process or team or software improvements that needed to be made before the next Nav ORT. Other flyby ORTs ranged from mission operations and instrument-team malfunction scenarios, called "green card exercises," to practice runs for the Deep Space Network critical operations at flyby, to the science team practicing spotting faint moons and rings in simulated imagery mimicking what it would be receiving on approach to Pluto. In all, across 2012–2014, more than forty project ORTs were designed, executed, and then forensically dissected in order to search for deficiencies that needed to be fixed, procedures that could be improved, or new kinds of training that was needed.

The ORT phase culminated in 2014 and early 2015 with three science team "encounter dress rehearsals" involving the whole science team; the APL, SwRI, and NASA media teams; as well as a group of six professional science communicators (dubbed "media embeds"), whom Alan had recruited to translate discoveries into real-time press releases, captioned image releases, and edgy "Pluto in a Minute" videos. In each of these, the science team worked with simulated Pluto images and spectra that John Spencer and a small cadre of others concocted (using, e.g., modified Cassini images of icy moons as stand-ins), replete with potential discoveries (like new moons, or puzzling surface features) for the team to practice on. No planetary mission had ever done this before.

Was all this really necessary? Alan felt that despite the deep experience of the individual players who had been on other teams for other missions, they had to practice as a team rather than just wing it in real time when New Horizons arrived at Pluto. He knew that there would be no excuse for mistakes or delays in any aspect of how Pluto and its moons would be revealed to the world by New Horizons in the summer of 2015.

The science team, many also busy on other space missions, agreed, but by the time the last of these massive science team ORTs was scheduled, with the spacecraft already on final approach in April of 2015, some science team members rolled their eyes at Alan's intensity. The first two science team ORTs had worked, and many lessons had been learned—as intended. Did they really need to go through all this yet another time? Was Alan becoming a space-age Ahab, relentlessly and obsessively searching for the great white whale of some hidden problem that might still sink the effort? Or was he a fearless leader pressing his team onward, onward across the interplanetary seas toward their ultimate victory? For some team members, it was hard to tell. But one thing was certain: no one would second-guess the thoroughness of preparations that the New Horizons team had made for "showtime" when it arrived.

PLANNING TO SHARE THE FLYBY

Flyby planning wasn't restricted to just mission science and engineering. One final piece of planning involved finding ways to maximize public engagement during the flyby. This is an area that NASA has a lot of experience in, so New Horizons didn't start from scratch. But Alan wanted to go big in this, and he created a vision for engaging the public at a level not seen since Apollo, a vision fit to commemorate the first exploration of a new, unexplored planet since Voyager went to Neptune in 1989.

This began with a formal NASA Communications Plan, designed and written by the New Horizons project in 2012 and 2013. To aide in this, workshops with writers, educators, social media experts, filmmakers, and science communicators were held to explore engagement themes and target audiences and then plan over two hundred communication efforts of all kinds. Then the project created videos, press briefings, printed materials, and even "Plutopalooza" party kits for schools and astronomy clubs. Then Alan recruited "influencers," like Bill Nye "the Science Guy," illusionist David Blaine, Queen guitarist Brian May (who is also a card-carrying Ph.D. astrophysicist), and other celebrities who were genuinely interested in New Horizons and wanted to help connect it with wide public audiences. The encounter promised some uniquely exciting moments of discovery and exploration. The educators, scientists, influencers, and communicators brought aboard for the flyby wanted to fully realize the potential to share the excitement and results of the impending flyby and to bring everyone who was interested along for the ride of a lifetime.

12

INTO UNKNOWN DANGER

FIVE'S A CROWD

Even as the encounter planning was well under way, the Pluto system was revealing itself to be more complex, and more crowded, than anyone had known. After Pluto's small moons Nix and Hydra were discovered back in 2005, the New Horizons project asked for and received more time to use the Hubble Space Telescope to undertake an intensive search for still more moons of Pluto. If there were other objects there to study, it was important to discover them far enough in advance to fit observations of them into the encounter sequences. Additionally, Pluto could possibly have rings, the New Horizons team surmised; these would also be important to know about in advance, both to plan their study during the flyby, and to avoid colliding with them.

It took years, until June of 2011, to find anything. But eventually planetary astronomer and moon/ring hunter extraordinaire Mark Showalter used the Hubble to take the deepest ever long exposures of the space around Pluto. Showalter was looking for faint rings, but in one of his super deep exposures he found a small, faint moon which turned out to be orbiting Pluto in between Nix and Hydra, circling the planet once every thirty-two days. The quadruple Pluto system had become a quintuple! Then, almost exactly a year later, in an even more sensitive Hubble search for faint rings, Showalter found yet another small moon, orbiting between Charon and Nix. The system had now become a sextuple! Both of Showalter's satellites were much fainter than either Nix or Hydra, and therefore likely to be much smaller, suggesting to many on the science team that still smaller moons might be there for New Horizons to discover when it got close.

For Showalter, finding two new moons was thrilling and interesting in its own right, but it also meant a greater likelihood of finding his elusive rings. As had been learned at the giant planets, small moons can generate thin rings by casting off debris when impactors strike them. Because very small moons have very low gravity, any debris blasted off them this way escapes into orbit around the moon's planet, getting spun into a ring.

As the fourth and fifth moons of Pluto, the two newest ones were at first just called P4 and P5. But in time, Showalter, working with Alan and others on the New Horizons team, as well as NASA, organized an online, public crowd-sourcing of ideas for naming the new moons. With the public's help nominating and voting on names, Styx (the goddess of the underworld river near Pluto) and Kerberos (the dog guarding Pluto's realm), became the official names of Pluto's fourth and fifth satellites.

A SPIDER'S BITE

With Showalter's discoveries of 2011 and 2012, it was becoming clear that Pluto possessed a rich system of moons, all orbitally interacting with one another. The emerging picture of a system thick with tiny moons and, potentially, thin rings was scientifically enticing, but it was also a spacecraft team's nightmare, implying a greater likelihood of debris laying in their chosen path through the Pluto system. After all, at the nearly ten miles per second that New Horizons would fly past Pluto, a collision with something smaller than even a rice grain could be catastrophic, striking the spacecraft with the energy of high-caliber ordnance and ending the mission immediately—even before its precious Pluto data could be transmitted home to Earth.

Alan did some so-called back-of-the-envelope calculations of possible ring density that could be produced by some undiscovered moons. If his simple estimates were right, then the spacecraft could be in real trouble at Pluto. He showed his results to the science team, and it got their attention, generating some urgency to look at this situation more closely with careful computer modeling. The fact is, the spacecraft was flying into the unknown. This of course is what made the mission so exciting in terms of the promise of new science and discovery, but it also meant they were flying into unknown danger.

Science collaborator Henry Throop created another model, and his calculations confirmed Alan's results. Glen Fountain remembers the reaction: "It scared the pants off of everyone. The bottom line was that we could be fatally impacted up to thirty times while going through the system."

Could it be that Pluto, the planet of their fascination and all their efforts over so many years, was actually a death trap for their spacecraft? As Alan put it to the team at one point, "What if the object of all our affection is really a black widow?"

So a concentrated effort to determine if the Pluto system might be hazardous for New Horizons began. This started with much more sophisticated computer modeling of what hazards might be there, and intensified to a long and thorough search for rings and moons or other orbiting debris as New Horizons approached Pluto.

Science team member John Spencer was assigned to lead this "hazards campaign," and he made it his baby. John got the job in part because of his deep expertise in both telescopic observing and spacecraft imaging techniques.

John and Alan mapped out a multistage effort to assess and mitigate the risk. The first step was to carefully reanalyze all existing data sets that could shed light on potential hazards. John recalls:

> First, we just needed to gather whatever other information we had to constrain the risk. That meant looking even more closely at existing Hubble images to see if there was any direct evidence of distributed debris around Pluto in the region of the moons, or if not, what limits we could set on them. Then we looked at data from stellar occultations. The Pluto community has observed many stars passing behind Pluto to study Pluto's atmosphere. Those stars would have also passed behind any narrow rings that might be in the system. And if there were rings, then as a star passes behind them, its brightness should dip. So we reexamined all those data sets to search for any evidence of faint rings.

Around the same time, Alan gave the spacecraft team the task of analyzing how well-protected New Horizons was from debris particle hits. The spacecraft was not defenseless: its body was covered in protective aluminum faceplates. But more important, its thermal blankets that overlaid those faceplates included layers of Kevlar shielding—the same material used in bulletproof vests. This had been done to protect New Horizons from interplanetary meteorite strikes as it crossed the solar system.

To assess how well those measures protected New Horizons, in 2012 and 2013 the spacecraft team used a special high-velocity gun to fire different kinds of particles at copies of these faceplates and blankets. The result was good news—the Kevlar shield was actually more effective in stopping impactors than design analysis had indicated. Using these findings, mechanical engineers at APL modeled the probabilities that particle impacts large enough to penetrate the Kevlar and spacecraft's aluminum skin could damage each given component of the spacecraft, every instrument, each fuel line, each electronics cable bundle, and each box of electronics. From this they derived much more detailed damage and destruction probabilities as a function of impactor size and velocity. The verdict: the lethal hazard concern was real.

BECOMING FAIL-SAFE

Because the possibility of lethal hazards at Pluto had become real, Alan wanted the New Horizons team to have something to show for all their work if the spacecraft was lost before it could complete the flyby and send its close-approach data to Earth.

The solution to this challenge was called the "fail-safe" data transmission. When he conceived it, Alan likened it to astronaut Neil Armstrong's "contingency sample" collection, the first thing Armstrong did after stepping onto the Moon in 1969. Then the logic was to have something to show scientifically for the mission of Apollo 11 in case something went immediately wrong, and he and Buzz Aldrin had to suddenly abandon their moonwalk before more comprehensive lunar sampling could take place.

Channeling the same logic, Alan asked Leslie's PEP team to draw up a list of imaging, spectroscopy, and other data sets that could be fit into a fail-safe data transmission to be sent back to Earth just hours before the flyby's closest approach, when the craft was still far enough from Pluto that there was no significant possibility of a lethal impact.

The fail-safe data wouldn't come close to substituting for the main data collection at Pluto, and it wouldn't prevent the mission's main objectives from being lost if a fatal debris strike occurred. But it would give them a sampling of the best data taken before closest approach to salve their wounds and learn as much as possible about Pluto and its moons if the spacecraft was soon thereafter destroyed.

But nothing is free, and adding the fail-safe transmission came at a price. Pointing the antenna back to Earth to send the fail-safe data home meant taking four precious hours out of the approach observations late on the scientifically crucial day prior to flyby. Some complained, but as mission PI, Alan's calculus was different:

> There just wasn't any way I was going to face NASA and the press if we lost New Horizons without having something important to show scientifically from the flyby. And I also wasn't going to put our team in the position of coming up completely empty if that

happened, and of saying we hadn't thought through the chance of failure and collected our own "contingency sample."

So the fail-safe data transmission became a part of the plan; as it later turned out, its best images were so good that they led the newspaper and web stories the day after flyby.

PLANNING AGAINST A BLACK SABBATH

The next part of the hazards effort was spent developing plans to use the onboard LORRI telescopic imager to look for rings or new moons as New Horizons approached Pluto. John Spencer:

> This imaging effort was planned to begin about sixty days out from Pluto, which is when LORRI became superior to Hubble for finding moons and rings. For seven weeks starting then, a series of hazard-search imaging campaigns was planned. Each campaign would make hundreds of images, which would be sent to Earth and "stacked" on the ground—combining many individual images in computers to make the most sensitive possible searches for faint moons or rings.
>
> As the hazard search images reached the ground, we planned to use software codes we built to look for any moons or rings. We also planned to build computer models to determine what orbits the debris would occupy and, in turn, what threat the debris on those orbits would pose to the spacecraft. From that we could determine whether any given hazard was acceptable or not.

And what if a given hazard found on approach was not an acceptable risk? Was there an alternative to simply plowing ahead, taking chances that fail-safe data was all the mission would get out of the twenty-six-year-long effort to explore the Pluto system?

There was. The approach, initiated by Alan, identified alternate trajectories through the Pluto system that would avoid various potential hazard zones. This meant planning several Pluto flybys—each with different paths through the system and different observation timing.

Recall the monumental amount of work, described in the previous chapter, which went into planning the flyby. Now the team would have to do that all again for each new pathway selected to dodge potential debris. It was a huge new workload and cost to the project, but given the lethality of debris strikes at thirty-five thousand miles per hour, and the reality of having only one spacecraft and no second chances, there didn't seem to be any other option.

For each of these new, backup flybys, the nine-day-long Core load of thousands of spacecraft and instrument commands that would direct the bird during its most intense and crucial flyby phase had to be completely redesigned, rebuilt, and fully retested.

Alan named these alternate trajectory plans SHBOT, for "Safe Haven Bail-Out Trajectory." It was pronounced like "Shabbat," the Hebrew word for the weekly Jewish day of rest. Alan isn't particularly religious, but he loved honoring the heritage of the many Jewish members of the team, including himself, Leslie, Cathy, and Hal, and he loved the way this word injected a little note of prayerful hopefulness as they prepared to confront the unknown hazards. Later, in response to a paranoid journalist's criticism that NASA was secretly planning to "bail out" of the flyby, Jim Green, the head of NASA's Planetary Science Division, asked that SHBOT be changed to the less threatening "Safe Haven by Other Trajectory," and it was.

The other aspect of SHBOT that was also meant to help protect New Horizons was the idea of changing the entire spacecraft pointing plan during these backup flybys. The Galileo and Cassini missions had both used a technique called "antenna-to-ram" to protect their spacecraft by turning to use their big dish antennas as forward shields when they crossed near the rings of Jupiter and Saturn, respectively. Flying in this orientation, most debris strikes would have to penetrate the dish antenna before reaching the Kevlar and spacecraft walls, providing an extra layer of protection. Tests performed with the high-velocity debris gun showed that the dish antenna on New Horizons could withstand many ring-particle impacts and still work, so this looked like a good way to add insurance against lethal impacts if the craft had to fly through possible debris. The engineering team showed that the antenna-to-ram protection provided a 300 percent reduction in the risk of a fatal debris strike to New Horizons.

Yet antenna-to-ram also introduced a major problem of its own. If New Horizons had to fly by with the antenna pinned in the direction of flight, it would cripple the spacecraft's ability to "point and shoot" in various directions to get observations of Pluto and its moons. So, although moving to the antenna-to-ram orientation would protect the craft from destructive impact, it would severely curtail the mission's ability to meet its science objectives. Scoring just how severely this damaged the flyby science was a task Leslie Young and her PEP team were assigned—grading the loss of observations for each SHBOT trajectory and comparing that to the flyby plan they had so intricately prepared and optimized.

The SHBOT trade-offs were painful, and the discussions about them were sometimes tense. John Spencer recalls some of the deliberations: "Many—maybe most—of the most important scientific observations were compromised or killed off if we had to choose the antenna-to-ram SHBOT. Obviously, people were very unhappy at that prospect. So there were lots of heated discussions about whether SHBOT was really a viable thing."

But Alan was adamant:

> I looked at it pretty coldly. Everything we had worked for to explore Pluto since 1989 came down to the success of the flyby. If the spacecraft was destroyed close to Pluto, we would not only lose all the observations after that point, we would lose all the data stored from the previous observations. If we lost our baby at flyby, almost none of that data would be sent back, only the fail-safe. As I saw it, in that case we'd get a near-zero for a grade—learning very little about Pluto and its moons. If we were really

faced with potentially catastrophic hazards, I was perfectly happy to trade the high grades of the optimal flyby for the low grades of a SHBOT flyby, because a dismal grade is a lot better than a zero. It wasn't what I wanted, but I wasn't going to blink if that was the only choice that had to be made.

13

ON APPROACH

HUNTING FOR STILL FURTHER DESTINATIONS

As the Pluto flyby approached, the New Horizons team intensified a search it began in 2011 to find a Kuiper Belt Object that the spacecraft could intercept and study after Pluto. This opportunity to study ancient bodies beyond Pluto—particularly small ones that were the building blocks of small planets like Pluto—was a key part of the Decadal Survey's motivation for its top endorsement of a Pluto Kuiper Belt mission back in 2003.

By 2013, John Spencer and Marc Buie, who were leading the search for this flyby target using the world's largest telescopes, had found numerous small KBOs but none that could be targeted within the fuel reach of New Horizons. With time running out before the 2015 Pluto flyby, the ground-based searches simply weren't yielding what was needed. So Alan decided to take a new approach. The primary difficulty was that the turbulence of the Earth's atmosphere blurred the myriad stars in the search images just enough to blend them with faint pinpoints of light of their potential target KBOs. The only way around this was to use the Hubble Space Telescope, which, because it orbited above Earth's atmosphere, offered the crisper view needed to separate still fainter KBOs from the dense fields of background stars.

John and Marc, working with Hal Weaver, calculated that a proper search with a high probability of success would require almost two hundred Hubble orbits, or about two continuous weeks of Hubble time—more than ten times a typical-size Hubble proposal. Such a large proposal would be a very tall order to win.

Making the challenge even steeper, the ticking clock to the Pluto flyby in 2015 meant the team could not go through the normal process of applying for Hubble time in spring of 2014, because that observing time wouldn't start until late summer 2014, too late given the positions of the Sun and the KBO search fields where the scientists had to look.

But there was opposition within the Hubble project to conducting such a large search on short notice. Despite that, the New Horizons team proposed the time it needed, with John leading the effort. Their proposal was turned down. They were incredulous. The Decadal Survey had directed New Horizons to study Kuiper Belt Objects.

Only Hubble could enable the New Horizons mission to undertake this study. Would NASA allow them to fail to have a Kuiper Belt mission after Pluto for want of a couple of weeks of Hubble time, 2 percent of Hubble's efforts that year? After all, there was no other credible way to explore KBOs anytime in the next several decades, except with New Horizons—and if they didn't get Hubble time, their mission was not going to have a Kuiper Belt flyby target.

Alan appealed to NASA Headquarters, and after a second John Spencer proposal to the Hubble project in the spring of 2014 and some high-stakes backroom maneuvering, the Hubble project announced that the KBO observing time for New Horizons had been approved.

Observations began the same week, under the pressure of time because the star fields that needed to be searched would be out of position as the Sun neared them in the sky by the fall. As the Hubble data began to rain down, several weeks of round-the-clock work to analyze the images, make candidate detections, and schedule follow-up confirmation observations began. John, Marc, and a cadre of post-docs and collaborators crammed months of search work into weeks, knowing that, because the Pluto flyby was looming and they would soon have to quit work on this, time was running out. One afternoon, Marc—who led the data-analysis effort—said to Alan and John, "You'd better come down to my office and look at this." He had found a KBO that New Horizons could reach!

Soon, Marc and his team found a second reachable KBO in the Hubble data, and then a possible third, along with several others that were close to, but not quite, reachable within their fuel supply. Follow-up observations confirmed that two of the three objects were indeed reachable.

The Hubble effort had succeeded; New Horizons now had a choice of two KBO targets to intercept for a flyby after Pluto! Both were just the kind of planet-building-block size they wanted, and both could be reached by early 2019, about three and a half years after the Pluto flyby.

ENTERING PLUTO SPACE

Occasionally, as New Horizons flew across the solar system, Alan would arrange for events or announcements designed to help engage the public, and to remind people of a spacecraft racing toward an un-seen world beyond the explored planets.

One of these events took place in 2008, as New Horizons was speeding away from Jupiter and beginning its long journey across the vast middle solar system toward Pluto. That October, a full-size rep-lica of the New Horizons spacecraft was "inducted" into the National Air and Space Museum near Dulles, Virginia, outside Washington, DC, and put on display. It was a rare privilege—fewer than 1 percent of all spacecraft are honored this way. At a public lecture commemo-rating the event, Alan announced that New Horizons was carrying nine mementos on its flight to Pluto and beyond. Each was symbolic.

1. A container holding a small portion of Clyde Tombaugh's ashes and the inscription Alan wrote about him.

2. A CD-ROM with more than 434,000 names of people who had participated in a "Send Your Name to Pluto" activity organized by the Planetary Society and NASA.

3. Another CD-ROM with pictures and notes from people on all the various teams who had designed, built, and launched New Horizons.

4. A Florida state quarter, for the state where New Horizons was launched.

5. A Maryland state quarter, for the state where New Horizons was built.

6. A small piece of carbon fiber from SpaceShipOne, which in 2004 became the first privately built piloted spaceship to reach space.

7. A small U.S. flag on the port side of the spacecraft.

8. A second small U.S. flag on the starboard side of the spacecraft.

9. That 1991 U.S. stamp proclaiming, PLUTO NOT YET EXPLORED, something New Horizons audaciously expected to make obsolete in 2015.

Alan ended that speech by saying that it was a privilege for the New Horizons team to carry each of these nine mementos, and he promised that once the first reconnaissance of Pluto was completed, the team planned to petition the U.S. Postal Service to issue a new stamp for Pluto, commemorating its exploration.

More opportunities for public engagement came during key milestones in the succeeding flight years, and these, along with popular articles, blogs, social media, and public talks, kept New Horizons in the public eye across the many years it took to reach Pluto's doorstep.

Then, in late summer 2014, a special occasion offered an opportunity to remind the public that the long journey was coming to an end and the exploration of Pluto was now less than a year away. New Horizons was crossing the orbit of Neptune. It was a powerfully symbolic moment with a clear message—"Next stop, Pluto!"

The emotional power of the Neptune crossing was heightened by a coincidence of dates that could not have been imagined when that small band of Plutophiles first began agitating for a mission to explore Pluto during the summer of 1989 as Voyager 2 explored Neptune: the New Horizons crossing of Neptune's orbit on August 25, 2014, came exactly twenty-five years to the very day after Voyager had swept past Neptune!

To Alan, the symbolism of that anniversary was too incredible to be ignored. So, working with NASA, the team created a public event both to commemorate Voyager and to build anticipation for the Pluto flyby, just over ten months away. As a part of the event, a panel discussion was held at NASA Headquarters in Washington, DC, and was streamed live on NASA TV for space fans all around the world. Moderated by this book's coauthor David Grinspoon, himself a veteran of Voyager from student and post-doc days, the panel featured New Horizons scientists Fran Bagenal, John Spencer, Jeff Moore, and Bonnie Burratti, each of whom had worked at the Voyager Neptune encounter. Each reminisced about the excitement and inspiration of Voyager at Neptune and how that exploration early in their careers had affected each of them. Now, in middle age, they were part of a team about to explore the next, even more distant planet for the first time.[*]

The panel's conversation eventually turned to careers and mentorship and how, just as their mentors had done for them in the 1980s, they were now themselves mentoring a new generation of young scientists who would be learning the ropes on New Horizons, and who would then, hopefully, be leading new missions of their own in the 2030s and 2040s.

[*]The video of this event can be seen at https://www.youtube.com/watch?v=DaUhaVUN3Yc

This made for a perfect segue. Alan called a group of young New Horizons scientists, many born during the Voyager era, to the stage. Like many in the public, they had never witnessed the intensity and excitement of a first planetary exploration, but they were soon going to learn firsthand just what it was like.

Next, Caltech's Ed Stone, the celebrated scientist who had led Voyager's scientific team since it launched, gave Alan an American flag that had hung in Voyager mission control, to now hang in New Horizons mission control. Within the New Horizons team, the poignancy of that event was further heightened by the fact that Tom Coughlin, the original New Horizons project manager, had passed away just two weeks before.

That day, as New Horizons crossed beyond the orbit of Neptune, a figurative baton had been passed: the banner of exploration of the solar system had been transferred from Voyager to New Horizons, and from one generation of scientists to another. Once it was beyond Neptune, New Horizons was now considered in "Pluto space," because it had reached the solar system's third zone, and was ready to soon explore it.

It was time for New Horizons to shine.

WAKING THE BIRD

The crossing of Neptune's orbit had been a milestone on Earth, but New Horizons slept right through it, gliding onward at its incredible speed in silent hibernation. Across the rest of 2014 too, New Horizons flew on in deep sleep, covering over 100 million more miles beyond Neptune's orbit by the time that December had come.

Back on Earth though, its team wasn't hibernating. Those months were a blur of final flyby simulations, planning for the expected onslaught of press and public attention, creating dozens of software tools to analyze soon-to-be-collected Pluto-system data, and the coding and testing of the early flyby approach sequences that would kick off almost as soon as 2015 did.

On December 6, 2014, right on schedule, New Horizons commanded itself out of hibernation for the final time on its long cruise out to Pluto. The spacecraft was now barely six months from its flyby. Alan recalls:

> Exiting hibernation for that last time meant showtime at Pluto was finally approaching. When we started hibernating back in 2007 after passing Jupiter, we found being in hibernation was pretty weird, because we'd been actively operating the spacecraft every day during the eighteen months since launch. It took us about a year to get comfortable and make hibernation routine. But by the time we got to 2014, we were so used to hibernation that it was like a warm blanket, and conversely the thought of living without it seemed weird.
>
> After all, we really hadn't operated the spacecraft day in and day out for more than a couple of months at a time in any given year since 2007. So the prospect of going back to that for all of the 2015 and 2016 flyby activities and the long downlink of data afterwards was a little daunting.
>
> But most of all, exiting hibernation for the final time in 2014 felt momentous, because it meant that the only thing in front of us now was the flyby itself. No more years to go. It was a turning of a page: we had crossed the entire solar system. We really were on Pluto's doorstep. 2014 was ending, and that surreal never-never year of 2015, which had been in our future for so long, was about to begin.

There was a gathering at the APL MOC to receive the signal from New Horizons, letting mission control know their bird had awakened for the flyby. NASA executives were there, along with a gaggle of reporters and camera crews. At the expected time, when the signal reached Earth—having traveled four hours at the speed of light from Pluto 3 billion miles away—Alice Bowman beamed and gave a thumbs-up. New Horizons was reporting in, ready for duty at Pluto! The room erupted in cheers; champagne, cake, and music ensued.

There is a long tradition in space missions to mark milestone occasions with a "wake-up song." That started all the way back in 1965 when the astronauts of Gemini 6 were woken up in flight with "Hello, Dolly!" and it continued across all the human spaceflights ever since. Somewhere in the 1990s, robotic missions began using music for milestone occasions, too. For the occasion of New Horizons emerging from its last hibernation en route to Pluto, Alan chose a piece called "Faith of the Heart," an emotional theme song from the TV series *Star Trek: Enterprise*. Its lyrics seemed so appropriate to the journey of New Horizons. In fact, when Alan heard this song, to him it seemed to be telling the whole story of the mission.*

The song begins "It's been a long road, getting from there to here." The poignant lyrics then tell a story about undertaking a long journey, overcoming adversity, and triumphing over opponents trying to hold you back, over many years, through having the perseverance to see your dream come true. But just how appropriate these lyrics were, the team members would not know until the following summer when they would discover the expansive heart-shaped feature that dominates the surface of Pluto. The song concludes with the lines "I can reach any star, I've got faith, faith of the heart."

ON PLUTO'S DOORSTEP

The flyby of Pluto formally began barely a month after hibernation ended, on January 15, 2015, when the spacecraft started to execute the first of its dozen or so distant-approach command loads, which ran until the beginning of April. Pluto was still just a dot in the distance—about 150 million miles away—and most of the scientific instruments aboard New Horizons could still not even detect it. But the flyby science was beginning with nearly round-the-clock measurements

---

*The lyrics are easy to find online and if you read the full song you'll see what we mean.

of the environment out near Pluto's orbit using the onboard plasma and dust instruments SWAP, PEPSSI, and SDC.

With the LORRI telescopic imager, New Horizons could begin to resolve Pluto and Charon as bright dots. A week's worth of LORRI images captured slightly more than one full mutual orbit of the pair around one another, and were looped together into movies. In them, Pluto doesn't stay put in the center of the frame with Charon merely circling around it. Instead, both are circling one another around an invisible balance point close to Pluto but between them. That back and forth yo-yo motion of Pluto and Charon was starkly different from movies of moons orbiting giant planets like Jupiter or Saturn, where the planet in the center remains fixed as if in bedrock—immovable.

There was something charming and engaging about seeing Pluto and Charon yo-yoing back and forth under one another's gravitational pull. No one had ever seen anything like it in all the decades of planetary exploration: a double planet doing a swing dance, with the larger dancer being pulled to and fro by its somewhat smaller partner. When NASA released a video of this mesmerizing orbital dance, it became a viral internet hit, its charm heightened even more so by its pixelated imperfections and the jumpy frames that were captioned "Not a simulation!"

By early April 2015, with New Horizons just over 100 million miles from the first binary planet ever explored, the planet was bright enough as seen from New Horizons that its Ralph color camera could detect Pluto and Charon for the first time. NASA released this first color image. It wasn't much—just two little smudges of light near one another. Pluto appeared distinctly larger and brighter and reddish, while Charon was smaller, darker, and obviously grayer in color. Again, not much—but the public reaction again went viral. Something about the novelty and authenticity of these images, the awareness that they were taken by a machine that humans had built and sent so far away, was exciting people across the world. Few in NASA or the New Horizons team realized, even then, that the growing wave of public excitement they were seeing would become full-blown Plutomania by July.

BACKSTAGE BEFORE THE MAIN EVENT

To the outside world, besides the occasional image releases, not much was seen happening aboard New Horizons or in the project, but behind the scenes the mission team buzzed with activity during the early months of 2015. It was like being at a fine restaurant just ahead of a big banquet. If you stepped inside in the afternoon, the dining room would look quiet and peaceful, as if nothing was happening. But back in the kitchen you'd see something completely different—a blur of frenzied activity as the chefs and staff rushed to be ready for the main event.

One key effort taking place backstage on New Horizons involved navigating the spacecraft to home in on Pluto. Those LORRI images of Pluto and Charon orbiting one another that went viral hadn't been shot just for PR purposes. They were part of the critical early approach "OpNav" campaigns. OpNav is space geek–speak for "optical navigation," imaging Pluto against star fields to very precisely determine what engine burns were needed to reach the exact aim point around which the team had designed the Core flyby sequence.

Because New Horizons could only point and shoot its close-approach images based on advance calculations of where it would be relative to Pluto and each of Pluto's moons, mission designers calculated that the spacecraft could be no more than sixty miles off target for the flyby of Pluto; in addition, it had to arrive within less than nine minutes of its appointed time. If New Horizons didn't navigate well enough to meet both of these criteria, its closest images would be blank or partial frames, and the flyby and the mission—the whole, crazy quarter-century of effort to explore Pluto—would devolve into failure.

During approach, the mission navigation teams (there were two) needed to measure how closely Pluto's actual position against the stars compared to carefully calculated predictions. Analysis of these navigation images allowed the nav teams to determine how much correction was needed through engine burns to thread the eye of the needle. The two different nav teams, making independent calculations, served as a check on one another; the stakes were too high to do anything less.

Each week, as New Horizons rushed onward toward Pluto, the two teams sat down with Alan and Glen and encounter mission manager Mark Holdridge to present their latest calculations. By early mid-February, it was clear that a first, small engine burn was needed to correct their course to the intended aim point. So Alice Bowman and her team designed, tested, and then sent the engine-firing commands up to New Horizons. On March 10 the spacecraft burned its thrusters for 93 seconds to tweak its approach speed by just 2.5 miles per hour. Not much, but that correction would eliminate the 7,000-plus mile error that the LORRI images indicated that New Horizons had been off course. The correction maneuver went off flawlessly—New Horizons was now aimed precisely at its intended bull's-eye!

IT BEGINS

In late May of 2015, Alan moved out to APL in Maryland for the flyby. This was it. He would not be back at home in Boulder for more than a few days until the flyby operations were complete at the end of July. By then, the decades-long goal of exploring Pluto would have either succeeded or failed. The moment of truth had arrived.

A few days later Alan's executive assistant, Cindy Conrad, also arrived at APL to begin getting the logistics in order for the onslaught of science, nav, instrument engineering, and public engagement teams soon to take up working residence there for the flyby. By the end of June, the on-site New Horizons team at APL had swelled to over 200 flight controllers, engineers, scientists, and others, working seven days a week, virtually around the clock. They took over a large building at APL's space department, with dozens of offices, team meeting rooms, break rooms, conference rooms, and even places for people to sleep. The sleeping rooms and cots were key because the workload was simply nonstop, day and night.

New Horizons had passed its "better than Hubble" imaging threshold back in April, and as a result, the LORRI camera was now starting to discern never-before-seen features on Pluto, and the science team was already beginning to discover new things about Pluto, even from tens of millions of miles away.

For example, it had been known from Hubble imaging that one hemisphere of Pluto—the one they had targeted to fly over at closest range—had a vast, bright, reflective area splashed across it. Spectra from Earth-based telescopes had revealed this place to be rich in nitrogen and carbon monoxide ices.

And as LORRI images of Pluto began to transform the planet from a dot in the distance into a disk with distinct, broad-scale markings, a continent-size trapezoidal feature appeared on this bright hemisphere. Because of its shape, Alan nicknamed the feature "India." Interpreting these early images was a bit like an inkblot test. Images of the "far side" of Pluto—the hemisphere that would not be seen at closest approach—revealed four very nearly equal-size and equally spaced dark regions across Pluto's equator that were nicknamed the "brass knuckles." Pluto, a point of light since Clyde Tombaugh discovered it 85 years before, was now, finally, becoming a real place, before our very eyes.

ENTER THE PRESS

During the final weeks as New Horizons swept down on Pluto, there began a tidal wave of requests for press interviews. Literally hundreds of press outlets, from magazines to newspapers to TV documentaries to television networks, requested background stories, asking such questions as "Why are you doing this?"; "What do you expect to find?"; "How did you become involved?"; and "What's your greatest concern?"

There was also great interest in the finer details of the flyby mechanics, the team members' personalities, the science of the Kuiper Belt, and much more. Nothing like this—the first exploration of a new planet—had occurred in a generation, and the media picked up on the historic nature of the events.

Knowing that they couldn't all get background "B-roll" footage at the flyby itself, dozens of network and documentary film companies—from across the United States, Canada, Europe, Australia, and Asia—wanted to shoot background interviews ahead of the flyby. The growing public interest was exhilarating for the New Horizons team. But the other side of that coin was that those scientists and engineers had to add multiple press interviews every day on top of all their other work. They had anticipated this would happen, but now, in the growing heat of the flyby, they found themselves taxed even more than expected. Alan:

> We were all working to get our technical job done, and our project management job, and our science job, and worrying about the spacecraft engineering, communications planning, navigation, and several hundred emails per day. But then, all of a sudden, there's a whole other layer to the workload: every day there were several hours of interviews, often stretching into the night. And groups of school kids were coming in, too; as were local officials, national politicians, science glitterati, and even a few celebrities; and all the while APL was holding dinners and staging events for their staff and the public. We were all working seventeen- and eighteen-hour days by then, every day.
>
> After a couple of weeks of this, I realized, "This is the new normal; I am simply not going to get more than four or five hours of sleep on any night until after the flyby. And many nights it's going to be less." But rather than resent the lack of sleep, I decided to feed on it for energy—to be carried by the momentum of the flyby across six sleepless weeks, powered only by adrenaline.

HAZARD WATCH—CLEARING THE PATH

In the midst of all this hubbub of public and press attention, a deadly serious activity was playing out on the spacecraft and back at APL: the "hazard watch" imaging campaign, to determine if the planned path through the Pluto system would be safe for the spacecraft to fly, or if instead they would need to divert at the last minute to a safer but less scientifically rewarding path through Pluto's thicket of moons.

For the next seven weeks, through late June, a series of four intensive hazard searches took place. Each consisted of the same steps: the New Horizons LORRI camera blanketed the space around Pluto with sensitive images designed to detect tiny satellites and incredibly faint rings that still could pose a hazard to flight. These imaging campaigns were followed by days of data transmission to get those images to Earth, which were in turn followed by days of careful scrutiny by a team of fifteen data analyzers using sophisticated image-analysis software packages to tease out even the faintest details from the hazard watch images. Finally, once the images had been analyzed, models were run and their results were presented to Alan and Glen giving the resultant probabilities of "loss of mission."

Right off the bat in early May during the first hazard-watch imaging campaign, expert satellite hunter Mark Showalter thought that his software had found a new moon—in the very first set of "better than Hubble" hazard-search images. Alan thought, "Oh, here we go. As soon as we looked, we found the first one. How many are there going to be?" But fortunately, upon more careful analysis, Showalter's "moon" turned out to be an artifact of the computer processing. That scare got everyone's attention and provided a sobering focus on the coming weeks of hazard searches. John Spencer:

> Our first big decision point about whether to stay the course or divert was at thirty-three days out from Pluto, because the least fuel-costly engine-burn trajectory deflection to a safer course could have been made about then. Then we had other decision points at twenty days out, and then a similar cycle a couple more times in the last weeks of approach. Our last deflection opportunity was fourteen days out from Pluto.

The verdict at each decision point: all clear; stay on target. No new moons, or rings, or other hazards were found. Once it became too late to burn the engines if hazards were found, they could still choose the "antenna-to-ram" pointing option, which would use the high-gain dish antenna aboard to shield New Horizons from impacts. John:

> We got our last set of hazard observations about thirteen days out. But still, no hazards were found. So at about eleven days out, we made the decision that there was nothing that warranted uploading the antenna-to-ram sequence instead of the nominal sequence. It was at that point that we felt we had really done our job. That's when we went out and celebrated.

To the best knowledge that New Horizons could provide, the path was clear! Of course the spacecraft was still at risk of being disabled by hazards that couldn't be detected, but months before, Alan and Glen had gotten an agreement with NASA that if there was no *known* hazard, New Horizons would stay the course and proceed on its planned pathway through the Pluto system.

As had the Voyagers, and as had every pioneer from the first humans who left Africa to populate the rest of the world, to the Vikings, to the Polynesians, to the Spanish and the Portuguese, to Amundsen and Shackleton, to Hillary and Norgay, to Yeager, Gagarin, Shepard, Glenn, Armstrong, and Aldrin: they were flying into the unknown in the service of exploration. Despite all their preparations, hazard searches, calculations, and computer models, New Horizons could not explore Pluto without some risk.

As June ended and July dawned, the mission was about to make headlines and history, but exactly which headlines and what history couldn't be known until the last hours of July 14, when New Horizons would or would not check back in after flying through the Pluto system and making its exit, hopefully safely.

A CERTAIN TREPIDATION

As we recounted earlier in this book, more than 2,500 Americans had worked to design, build, launch, and fly New Horizons. As July of 2015 began, emails and phone calls from those engineers, technicians, launch crew, and others were pouring in to the flyby team as Pluto approached, urging the flight team on, saying over and over again: "You did it! We did it! We're finally there, go get 'em!"

For Alan, Alice, Glen, Fran, Leslie, John, Jeff, Bill, Marc, Chris, and so many others, the Pluto flyby had been a major organizing principle of their lives—some, for half or more of their careers. Now, each day, in images beamed to Earth from the Kuiper Belt, Pluto was growing in size. And it would continue to grow, day by day, until, in almost no time at all, it would be in their rearview mirror, and then, it would recede.

What would life be like after the flyby . . . ? One evening about two weeks out from the flyby, Alan went for a walk around the lake by his hotel with Amy Teitel, a historian and journalist, and one of the "media embeds" he had hired to help translate flyby science into NASA press releases. As they circled the lake, their discussion turned to his state of mind. Alan:

> Amy caught me off guard a little. She said something like "In a few days, the biggest thing in your career is going to happen, and then it's going to be over. Nothing you do later will possibly equal it. Can you cope with that?" And then she said, "A lot of people might face a nervous breakdown once something like that is behind them. How will you handle it?"

Many scientists and engineers on NASA flight projects experience some version of what Amy was talking about: they focus their efforts so intensely to achieve mission success, bonding with their team in doing something larger than life, that then, as the pivotal moments approach—the launch, reaching the target, the landing—they begin to feel a dread within. They foresee the energy and common purpose of the project evaporating; they see their team dissipating; they dread the loss of a long-sought, difficult goal no longer being in their future. In a way it's like graduation day, when you know your life as it has been, with all its purpose and social fabric, is soon to be ending for an uncertain plunge into a new future.

Others working on New Horizons were feeling it too. A few weeks before the flyby, Alice Bowman came to Alan and said that some on her mission operations team were dreading the flyby that they had worked so many years to plan. They wished they could put the brakes on it, she said, and savor this time and place a little more. Some, she said, didn't quite know what they would do when the beacon of exploring the farthest planet ever attempted was in their past, rather than their future. Alan:

> When Alice told me that, I realized that for so many years exploring Pluto had been a bright shining light, beckoning all of us onward to the future and I thought, "God, we're all feeling the same thing."
>
> I told Alice to tell her folks to look forward to having actually accomplished the exploration of Pluto, something almost no one thought our team could pull off when we started. I told her to tell them that they could revel in the images and other data, and all we were about to learn about this entirely new world. And I told her to savor every day left from here to Pluto, because none of us would probably experience anything quite like this ever again.

14

JULY 4TH FIREWORKS

CORE LOAD COLLAPSE

The weekend of the Fourth of July holiday gave many on the New Horizons team a brief but much-needed break and a last chance to recharge their batteries before the upcoming flyby, then just ten days away. Mission operations remained on duty to fly the spacecraft, but most others took time off to fire up grills and relax. It was reminiscent of the Christmas break the launch crews had taken in late 2005, which helped morale leading into the nonstop intensity of the launch campaign that following January.

Long before dawn on the morning of the Fourth, flight controllers in the MOC were to radio the Core load up to the spacecraft. This was the long command script that New Horizons would follow to execute its many hundreds of scientific observations during the nine days surrounding closest approach and flyby. Aptly named, this exhaustively tested command script would literally perform the core of the mission, and its faithful transmission and execution would direct New Horizons through every twist and turn, every computer memory assignment, every communication with Earth, every camera shot, and so forth.

The uplink was Alice Bowman's show, and Alan wanted to witness it. He considered seeing the Core sequence launched by radio commands to New Horizons akin to seeing the spacecraft launch first-hand. So, following his habit of wanting to be in mission control during the most crucial moments, Alan arrived at the MOC at APL at about three thirty in the morning to watch the flight controllers send the Core load. No one was there except for Alan and the two-person flight-control team. He brought them doughnuts, a familiar good-luck tradition.

After arriving, he sat in the back of the darkened MOC for about an hour and a half just watching and making occasional small talk while the flight controllers sent the command load from Earth out toward New Horizons in the Kuiper Belt. He thought a lot about all the years it had taken to make this happen, how much was riding on the successful operation of this single command load, and how much new knowledge this load would create. As he watched, he felt proud of the New Horizons team for a decade-long flight mission done so well.

By 5:00 A.M. the entire flyby command script was on its way to Pluto, ripping across 3 billion miles of yawning vacuum at the speed of light. Satisfied that all was well, Alan went to his APL office to get some work done. With almost everyone gone for the Independence Day holiday, it was a chance to get caught up on the growing torrent of email, meet with the public-affairs and encounter-logistics teams, and do a couple of phone interviews with the press before the onslaught of the next week began.

Among the emails waiting in his in-box were two messages from Alice Bowman that had come overnight. The first read, "Please don't go to the Mission Operations Center during the command upload. If the PI is there, it'll distract the flight controllers during this critical operation." Her second message read, "I know you want to be there at this special time, but this is one of my superstitions. I just feel like we should let them work this alone. Even if you're just somewhere in the back of the MOC, I'm worried it'll be bad luck." Alan regretted not having seen these emails earlier but it was too late now, and besides, everything had gone fine in the MOC.

As the morning passed, Alan thought more than once about the flyby command load winging its way across the solar system to New Horizons:

> I kept watching the clock, thinking it's been an hour now—the load has crossed the orbit of Saturn; it's been two hours now—the load has crossed the orbit of Uranus. By mid-morning, the load had reached the spacecraft out near Pluto, covering in just four and a half hours the distance New Horizons had taken nine and a half years to traverse. In another four and a half hours, I thought, we'll get the signal back that confirms the load has been received and properly loaded into memory. I went back to work.

LOSS OF SIGNAL

Early that afternoon, Alice Bowman was in the MOC with a handful of other mission operations personnel, waiting to see the report come back from New Horizons indicating it had received and stored the Core load. At about 1:00 P.M., and right on time, the first signals started coming back confirming the reception of the command script. Alice:

> Everything was going fine until we hit about 1:55 in the afternoon. Suddenly, we lost all communication with the spacecraft. Dead silence. Nothing. We'd lost comm. And it didn't come back.
>
> Nine times out of ten, when we lose signal, it's a problem with the ground station: something's out of configuration, or whatever. Because this upload was so important, we had our network operations engineers online. We call them NOPEs: that's their acronym. We also had our Pluto Aces—which are the controllers there in our ops center. So we had the Pluto Aces ask the NOPEs at the ground station in Australia to check their system configuration. All those checks came back that everything was nominal with the ground system.

That meant that the problem was not down here on Earth—not in Maryland where Alice and her team of Pluto Aces were gathered, nor in Australia where the NOPEs were at the Canberra station of the Deep Space Network receiving the signal from New Horizons. The loss of signal was due to a problem with the spacecraft itself.

Loss of communications is about as bad a thing as a mission control team can experience—it means the link to Earth is broken. But that's not the worst of it. It could mean the spacecraft had suffered a catastrophic failure. Alice felt a ripple of unfamiliar fear:

> You know that feeling in the pit of your stomach when something is occurring, and you can't believe it's happening? We'd come nine and a half years on this journey, and I couldn't believe this—we'd never lost communications. You allow yourself that five, ten seconds of feeling that fear and disbelief, but then everything we trained for started to kick in.

The sudden loss of signal fed the worst fears that something catastrophic might have happened to the spacecraft. New Horizons was still millions of miles from Pluto, and any hazards it posed. The chances of striking anything there in interplanetary space were absurdly low. But, nonetheless, everyone on the team had the passing nightmare thought: Could we have just hit something? Glen Fountain recalls:

> I was home when Alice Bowman called me and said, "We just lost contact." I only live about ten minutes away from work, and I made it back to the Laboratory in record time. Driving back into the Lab, all kinds of things go through your mind. I called Alan, and he was at APL, so he actually beat me to the MOC.

When Alan got the call from Glen it was surreal. He couldn't believe he was actually hearing Glen Fountain telling him, his voice laden with gravitas, "We have lost contact with the spacecraft." It was a frighteningly serious problem. Alan:

For a second I thought, "Alice's premonition about my being in the MOC this morning when we radiated up the Core load is coming true." Of course, that's completely illogical, but I did have that fleeting thought.

But I put that out of my head. I was out of my office and into my car within ninety seconds, driving the half mile or so over to the building where the MOC is. On the way I called NASA Headquarters to give them a heads-up. I parked and ran through building security and into the MOC.

Because the team had telemetry from the spacecraft before it went silent, Chris Hersman and his engineers, already arriving as well, had some clues to work with. Something key they discovered very quickly was that just before the spacecraft's signal stopped, the main computer had been doing two things at once, both of which were computationally demanding. One of these tasks was compressing sixty-three Pluto images taken previously, in order to free up memory space for the close flyby imaging soon to begin. At the same time, the computer was also receiving the Core load from Earth and storing it in its memory. Could the computer have become overloaded by this intense combination of computational tasks, and as a result rebooted?

This was Brian Bauer's theory. Brian was then the mission's autonomy system engineer, who had coded the recovery procedure that the spacecraft would automatically go through in just this situation. Brian told Alice, "If that is what happened, then the spacecraft will restart using the backup computer, and sixty to ninety minutes from now we'll get a radio signal with New Horizons operating on the backup computer."

The engineers, the Aces, along with Alice, Glen, and Alan waited out those long minutes, making contingency plans in case Brian's hypothesis was incorrect. But sure enough, after ninety minutes, a signal arrived from New Horizons indicating it had switched to the backup computer.

Communication had been restored, and with that, the fear of a catastrophic loss of the spacecraft evaporated. But the crisis wasn't over; it had just entered a new phase.

ONCE AGAIN: "WHATEVER IT TAKES"

The MOC and its surrounding offices were quickly filling up with engineers, more flight-control team members, and others on the project who had cut short their holiday weekend to come in and assist. People were arriving in shorts and flip-flops, in their picnic clothes, having dropped everything to get to the MOC.

As more telemetry came back from the bird, they learned that all of the command files for the flyby that had been uploaded to the main computer had been erased when the spacecraft rebooted to the backup computer. This meant that the Core flyby sequence sent that morning would have to be reloaded. But worse, numerous supporting files needed to run the Core sequence, some of which had been loaded as far back as December, would also need to be sent again. Alice recalls, "We had never recovered from this kind of anomaly before. The question was, could we do it in time to start the flyby sequence, scheduled to begin on July 7?"

That meant the team had just three days to put Humpty Dumpty back together again, from 3 billion miles away. If they couldn't, then with every passing day they would lose dozens of unique, close-up Pluto system observations that were part of the exquisitely constructed Core load flyby plans. The mission team suddenly found itself in a three-day race to salvage everything they had spent years planning and months uploading.

The New Horizons process to get back on track after any spacecraft anomaly is shaped around a series of formal meetings called ARBs, or Anomaly Review Boards. Soon after 4:00 P.M., only forty-five minutes after spacecraft recontact, the July 4 anomaly's first ARB was convened in the meeting room adjacent to the MOC.

At that kickoff ARB meeting the team members had to assess what had happened, how to restore the flyby plan, and how to make sure they wouldn't accidentally do anything during the recovery that would cause another problem on the spacecraft. The scope of how far they had been set back by the reboot onto the backup computer was stupefying. It was quickly estimated that they would have to perform the equivalent of several weeks of work in just three days to start the flyby Core sequence on time on July 7. And it would all have to be done flawlessly.

What made this even worse, was that every move had to be done by remote control with a nine-hour round-trip radio communication time between mission control and the spacecraft. Science classes teach how the speed of light is incredibly fast, how a signal moving at that speed can travel around the world in an eighth of a second or to the Moon and back—a half-million-mile trip—in just two and a half seconds. But for the New Horizons team trying to get their spacecraft back on track as it closed in on Pluto, the great distance between Earth and New Horizons made the speed of light seem excruciatingly slow.

Those assembling for the ARB knew that with all the press attention, the world would soon be aware that New Horizons had tripped over itself on the verge of its flyby. In just ten days, the spacecraft would hurtle through the Pluto system—nothing could stop that celestial mechanics—but whether it would be gathering the data it had journeyed almost a decade to collect, was something else.

Alan and Glen opened the meeting, telling the ARB that there was no finer spacecraft team they'd ever known than on New Horizons, and that if any team could pull off this recovery, it was the group in that room. Then Alice took the floor and began architecting how they would effect a recovery.

Alice immediately asked Alan about the science observations being lost that day and in the next three days before the close flyby sequence was to kick off on July 7. She wanted to know, from the PI, if her team should also attempt to recover those observations, in addition to reconfiguring the spacecraft and getting all the files and command load up to the bird for the close flyby. Alan:

> I didn't call for any discussion of it from the other science team members in the room. I didn't even let my flyby-planning czar, Leslie, weigh in. I knew for a fact that Alice's team needed crisp direction, with no fuzz on it, and that they needed to focus on

saving the main event, rather than the preliminary observations we were losing with the spacecraft idled due to the reboot. I told Alice that anything beyond getting us back on track to initiate the close flyby itself, on time, would be a distraction.

Alice wanted further clarification, and asked me, very precisely, "How much of the current command load's science can I trash?" I knew what was at stake. I knew what was icing and what was cake. I estimated that the Core load probably contained 95 percent of everything we wanted to accomplish at Pluto. All the other command loads combined, including this one which was now suspended because of the anomaly, were just details by comparison. I looked Alice right in the eye and said, "The Core load is all that matters to me. So do whatever it takes to kick it off successfully on the seventh. Trash as much as you need to in between."

With that, Alice had her marching orders. Her sole job now was to save the Core flyby sequence; everything else was expendable. But could it be done in time?

Alice and her team quickly but methodically devised a recovery plan. In the next three days, they had to design and build all the command procedures to get the spacecraft back onto its primary computer, then to resend all the lost command and support files for the Core load, and they would have to test all of this on the NHOPS spacecraft simulator before any actions were taken, to ensure that each step would work on the first try—there was no leeway for repeats. They knew when the flyby Core sequence needed to engage, which would be noontime on July 7. So Alice's team took the total time available until then and divided it up into nine-hour round-trip lighttimes—the amount of time it would take to send each set of procedures to run on the spacecraft and receive confirmation that it had performed successfully. Counting everything else that had to be done on the ground, they found there was time for only three of these communications cycles before the Core load would need to engage midday on July 7.

Thus, the recovery would be split into three steps. First, the team would command the spacecraft to restore normal, rather than emergency, communications. That would up the communications bit rates by a factor of one hundred, making the rest of the recovery possible to do in time. That first step alone, they estimated, would take about half a day to code, test, send to New Horizons, and get confirmation back that it had succeeded. Tick, tock.

Next, the team would command the spacecraft to reboot onto its primary computer. This was needed in order to use the flyby command load as coded. A reboot from the backup to the prime computer had never been done in flight. So a procedure had to be designed and coded for that, and tested on NHOPS, and the test results then had to be checked before that procedure could be sent to New Horizons. Finally, the team would have to methodically restore all the Core flyby files and engage the flyby timeline. It was nearly midnight by the time this plan was architected, and there was no time to spare: the clock had already bled down over ten hours since the loss of contact that afternoon. Tick, tock.

Alice's mission operations team, working closely with Chris Hersman's spacecraft systems team, wrote, tested, and then sent up the first set of commands about twelve hours after they had reestablished spacecraft comm, at about 3:15 A.M. on July 5.

Nine hours later, midday on the fifth, the MOC received confirmation that normal communications had been restored! But a day had passed, and New Horizons had swept nearly another million miles toward its destiny at Pluto. Recovery step 1 was complete, but now only two days remained until the Core flyby sequence needed to engage. Tick, tock.

THE INCREDIBLES

The New Horizons team organized their work, and their lives, for the next few days around the nine-hour communications cycles to the spacecraft and back. They ran on very little sleep and lots of adrenaline. They had worked together for over a decade and had encountered problems on the spacecraft before, but no problem of this scope or with such high stakes had ever occurred. It demanded a round-the-clock existence in mission control, and the team delivered.

Glen recalls, "The team just did what they needed to do. I started searching for places for people to sleep, trying to find something more comfortable than their office floors." And Alice remembers, "We found cots, blankets, and pillows, and someone brought in an air mattress. There weren't enough, so we were sharing." Alan:

> You should have seen it. Without a single complaint, people worked day and night—without so much as changes of clothes or places to properly sleep or shower, in some cases for four days straight. Some people were sleeping on desks. Some were living on just two or three hours of catnaps per day. There was no time for restaurant meals. We brought in people just to find takeout and keep the team fed.

RECOVERY

In order to ensure that this and every step of the recovery was going to work as intended, it was essential that each of the recovery procedures be tested on NHOPS. Because NHOPS so faithfully simulated the spacecraft, command-load testing on it could be used to work out bugs and certify that the instructions that would be sent to New Horizons itself would be error free.

As it turned out, a decision made years earlier proved to be a life-saver during the recovery. Recall that Alan had become so concerned that the team did not have a fully complete backup to NHOPS, that a second one was built. Well, during the weekend of July 4, there simply was not enough time to test all the new command loads needed to recover using only a single NHOPS. So they doubled up, using that second NHOPS to fit in more test runs. Had there been no NHOPS-2, the recovery would have taken days longer, and whole swaths of unique Pluto science would have been lost forever.

Using procedures tested on NHOPS-1 and NHOPS-2, the middle

step of taking New Horizons out of safe mode and getting it back on the primary flight computer succeeded and was confirmed by telemetry sent by the spacecraft on July 6.

Next, the spacecraft had to be configured just as it had been prior to the attempt to upload the flyby script on July 4, and then, as a final step, the Core load had to be sent back up again, and with it all the dozens of associated support files that had been lost when the anomaly rebooted the primary computer. Those steps and all the NHOPS testing for them, including many Anomaly Review Board meetings to plan and certify each step, took round-the-clock work on the sixth.

But somehow, by late morning on July 7 all the recovery work was complete. Exhausted, the team had managed to get the spacecraft back on track and ready to go for the flyby. They had completed it with just four hours to spare before the Core load needed to engage.

DOING THE FORENSICS

What science was lost because of the July 4th anomaly and recovery fireworks? In saving the day for New Horizons, Alice and her team did follow Alan's directive to do "whatever it takes" to save the Core flyby. So in the end they did trash all the observations that would have taken place during the three days of the anomaly recovery, because there was simply no way to replan them and also get the spacecraft out of safe mode and ready to start the close flyby on time.

But Alice's team did manage to save the sixty-three images that were in the process of being compressed when the anomaly occurred. Those images had to be compressed to fit in storage because the larger, raw images had to be deleted to open up more memory space for flyby data. During the recovery operations, Alice's team spotted an open window in the spacecraft operations timeline and managed to get that compression rescheduled, saving every single one of those precious sixty-three images.

What about all the approach observations that were trashed during the July 4 weekend recovery of the spacecraft? Alan assigned flyby planning czar Leslie Young the task of forming a tiger team to analyze just that. Leslie and her troops worked during the three days of the spacecraft recovery to look at every lost observation and its impact on the overall science return at Pluto. They found that each one had a later observation that was at higher resolution or closer range, meaning no objectives had been lost—except in one case. That was the final satellite search around Pluto that had been planned to take place on July 5 and 6, when New Horizons was still far enough out to blanket the space around Pluto with images. That sequence would have searched with several times the sensitivity of the previous search made just days before the anomaly occurred. When all the satellite search images were later scoured carefully by the New Horizons science team, no new satellites were found. This surprised many on the science team, since every time the Hubble Space Telescope had looked harder, it had found more moons. Would New Horizons have discovered satellites in that trashed, final, better search? No one knows, or will know, perhaps, until some future Pluto orbiter mission arrives, to search again.

And why did the July 4 anomaly happen in the first place? Shouldn't the team have anticipated the combination of activities that led the main computer to be overwhelmed, and tested for it?

The sequence running on New Horizons on July 4, had been thoroughly tested. But as it turns out, the activity overlap that produced the computer overload only happened because of a fluke in timing in the way the DSN transmission to load the Core sequence was scheduled and executed. Had the Core load been sent just hours earlier or later, the computer would not have needed to store it while also doing the intensive work of compressing those precious Pluto images. So, should the team have realized that these activities could have overlapped and specifically tested for that possibility? In hindsight, yes. But when the intensive load testing for the flyby sequences took place back in 2013, the DSN schedule for 2015 wasn't set, and the probability of that bad confluence between Core load storage and image compression occurring was very small. Alan:

Looking back in hindsight, there is no doubt that we should have spotted the possibility of the bad confluence and tested for it as a contingency, if not in 2013 before the DSN schedule for Core load transmission was set, then in 2015 after it had been. That oversight is on us, and it caused our Fourth of July fireworks. But it's a wonder to me that that was the only detail we missed— among literally tens of thousands in the encounter operations— that marred any portion of the flyby. All those years of planning, testing, simulating, asking so many what-if questions, and more, really paid off, creating a bulletproof flyby plan in every other respect.

15

SHOWTIME

WHEN BETTER ISN'T BETTER

Amid the many dozens of activities going on behind the scenes at APL as the close flyby was beginning, a key decision had to be made: as we previously explained, to accomplish the objectives of the flyby, the spacecraft had to arrive at the closest-approach point within plus or minus nine minutes—just 540 seconds—of the planned time. Only then could all of the spacecraft pointing maneuvers properly center Pluto and its moons in the camera and spectrometer boresights.

Much of how New Horizons achieved this goal was done with careful optical navigation and rocket-engine firings that the space-craft performed to home in on the closest-approach time. But mathematical analysis had shown that this alone might not be good enough to guarantee arrival in the critical plus-or-minus-540-second window. So the spacecraft's designers at APL also built in some clever software to correct for any remaining timing errors once it was too late to fire the engines. That software is called a "timing knowledge update"; what it does is adjust the onboard clock on New Horizons, basically faking out the spacecraft to think that it is a little bit ahead of or behind where it really is in executing the Core load. The end result slides all the planned flyby activities backward or forward by up to 540 seconds to synch them up with the final, best predicted time of arrival. The process to do this had been tested many times in ground simulations using NHOPS. But it hadn't been needed at the Jupiter flyby in 2007, so it remained something that had not been proven aboard New Horizons itself.

As the spacecraft approached Pluto, every day the optical navigation team used new images to determine just how far off the closest approach timing was going to be, and then calculated the timing knowledge update needed to correct for that. Concurrently, Leslie Young and her encounter planning team used sophisticated software tools to generate a "science consequences report," in which each close-approach observation was simulated for the newly predicted timing error to determine, assuming no correction was made, which would succeed and which would fail.

Remarkably, once the final engine burn had been made and New Horizons was on final approach, the predicted timing errors were looking surprisingly small—less than two minutes—way inside the nine-minute-long maximum acceptable error. Leslie's science consequences reports showed that there was no observation predicted to fail if no correction was made, though a few observations would be improved— by being better centered—if the team did make the timing knowledge update correction.

Now, with just days to go before the flyby, it was time to decide how big a timing update to radio up to New Horizons. The meeting to make this decision was structured so that at the end of all the discussion, after all of the navigation calculations were reviewed and the consequence report results were shown, after everybody on the team had asked all their hard questions and probed at every what-if, Mark Holdridge, who rode herd on this process, would go around the room, asking each technical lead if he or she was go or no go for the timing update. That polling started with the system engineers and went through mission operations, then navigation, then the project scientists, and then to project manager, Glen Fountain. As the

PI, and the final arbiter, Alan would be last, making the ultimate decision. Alan:

> Mark went around the room, and everybody was giving "Goes."
> I was pretty surprised. We were well inside the nine-minute box
> we had to be in, and while Leslie's report showed that a timing
> update could make a few observations work better, none were
> going to fail. I knew that our team wanted to score as high a
> grade as it could on every observation, but from my perspective,
> we were already lined up to get a straight-A report card. Fur-
> ther, the timing knowledge update was untested on the space-
> craft. I was surprised that no one was asking, "Was the risk
> of making things worse if the timing update didn't work as
> planned, worth the tiny gains in score it offered?" The poll got
> all the way up to Glen Fountain, who as project manager was
> just before me in the pecking order, and he too gave a "Go."
> I couldn't believe it! After all, we had just been through a "near
> death" experience days before, in which we'd overlooked some-
> thing subtle and sent the spacecraft into safe mode over the
> Fourth of July weekend. So, I was a little incredulous that every-
> one wanted to do this, even Glen, and that no one was thinking
> that, well, "better" might be the enemy of "good enough."
>
> In my mind, the situation was analogous to our second launch
> attempt in Florida back in 2006 when everybody said "Go"
> and I had to stop it, because I didn't want to have the con-
> trol center at APL operating on backup power during launch.
> I didn't like doing it, but I derailed that entire launch attempt
> because I didn't want to take a risk that everyone else had
> accepted.
>
> Just like that 2006 launch poll, the timing update poll had
> been unanimous until it got to me, but I said, "No Go," and I
> explained my reasoning. So I asked the team, "Am I missing some-
> thing? Is there a must-do reason to try out the timing knowledge
> update when we're already way inside the timing box?" No one

pushed back, even when I asked twice. So I vetoed the timing knowledge update.

After we ended the meeting I went back to my office on the other side of the APL campus and found I'd already received a number of emails and other messages from some who'd been there. Every single one of them expressed relief that I'd resisted group-think, and made that call, avoiding "better" for good enough. I was glad, too. There were still a hundred ways the flyby could fail—we didn't need this unnecessary one.

A MAELSTROM IN MARYLAND

As the spacecraft barreled onward toward Pluto, press coverage on television, in newspapers, and on the internet was building rapidly. So were the crowds at APL.

The press and public activities there were held in the Kossiakoff Center, a building at the front of the APL campus, with a large auditorium, a complex of offices and meeting rooms for press activities, briefings, and interviews, and an open space for large crowds of visitors. As people arrived, they were given different-colored badges to differentiate among press, visitors, VIPs, and staff. During the week of final approach, hundreds of reporters, correspondents, and documentarians arrived, from media outlets in North and South America, Asia, Europe, Australia, and Africa.

But it wasn't just press and friends and family coming to APL. In mid-July 2015, the Applied Physics Laboratory in itty-bitty Laurel, Maryland, became the *only* place to be for hard-core space freaks: there was simply no way to miss being present for the first flyby of a new planet in a generation. The internet made it possible to learn what was going on, to see new images, and enjoy the building Plutomania from almost anywhere on Earth. But there is still something irreplaceable about human beings gathered together for common purpose, to experience something larger than life.

The public relations team at APL had managed large public events before, but it was caught off guard by the massive and ever-growing outpouring of interest in New Horizons. By the morning of July 14, the day when New Horizons reached Pluto, the crowd there was more than two thousand strong, the phones in the press areas were ringing off the hook, and the mission and NASA web servers were churning furiously with hundreds of millions—and eventually billions—of visits from people tuning in from every continent of Earth.

The crowd at APL included a who's who of planetary exploration and many pivotal figures in the exploration of Pluto. Alan's old professor Larry Esposito, whose POSSE proposal had narrowly lost out to New Horizons fourteen years earlier, was there. In an alternate reality this could have been his flyby, but there he was, smiling and sharing in the excitement. Original JPL Pluto mission study leads Rob Staehle and Stacy Weinstein were there, too. And so were key people who had played decisive roles in making New Horizons happen, such as Mike Griffin, who had been APL's space department director during the late-stage construction of New Horizons and then went on to be NASA Administrator during the run-up to launch and the early years of the mission. Members of the current NASA brass were also out in force, such as former space shuttle commander Charles Bolden, who was then serving as NASA's Administrator. Bolden took special delight in the company of the "Pluto Pals"—an excited gaggle of nine-year-olds who were all born on the day of the New Horizons launch, who had been invited to join in the thrill of the flyby.

Celebrity sightings added something more, something almost surreal to the atmosphere at APL. Among the luminaries who visited were Bill Nye, Senator Barbara Mikulski, Queen guitarist Brian May, illusionist David Blaine, and the rock band Styx.

All the while, of course, the New Horizons team members were working behind the scenes to carry out the flyby. Recognizable in their crisp black team polo shirts with the New Horizons mission patch on the chest and a small American flag on one shoulder, many found themselves mobbed for interviews, autographs, handshakes, and selfies with bystanders each time they emerged from backroom meetings.

PLUTO'S HEART

In the week before the close flyby, years of speculation began to melt away as Pluto revealed more and more of itself, growing larger and clearer in the successive images sent back to Earth. We had long known, since the Hubble images and Marc Buie's deft construction of a crude surface map created from them many years before New Horizons arrived, that Pluto's surface was varied, with sharp contrasts between bright and dark areas. But there had always been the possibility that these diverse features were merely "painted on" the surface.

As team scientists began to recognize patterns in the images, they did what humans have always done: they started to give names to things. These first names, of course, were inherently temporary, as Pluto's true appearance was being clarified with every passing day. And because these names felt disposable, they were often whimsical (like the equatorial "brass knuckles" described earlier). One dark elongated region appeared near the equator, shaped vaguely like a cartoon whale, and that's what it was called, the "whale." Then the flyby hemisphere's brightest and largest spot, which Alan had first called "India," again rotated into view, appearing much larger than the last time it had been seen, one 6.4-day-long Pluto rotation before. Now it appeared rounded, with a double-lobed northern section, and in the south it tapered toward a sharp point. NASA press liaison Laurie Cantillo saw it and immediately asked, "Does anybody else notice that bright feature has the shape of a heart?" and once she said it, no one could get it out of their minds. It did look just like a heart!

The next day NASA made a press release announcing "Pluto has a heart," which promptly went viral. There couldn't have been a more perfect hook for even greater public engagement as the meme of Pluto's heart soared in social media trends. It became the iconic feature of Pluto, creating an emotional attachment for this small, previously indistinct planet at the edge of our planetary system. Within just days, the "heart" became enshrined on countless internet cartoons, on T-shirts, on dresses, on refrigerator magnets, on pieces of custom jewelry, and on children's plush toys. As Laurie Cantillo later put it, "In the summer of 2015, the world 'hearted' Pluto."

"I COULDN'T TAKE MY EYES OFF IT"

Near midnight on Monday, July 13, NASA's Deep Space Network of communications antennas received a precious package from New Horizons: the last and best data that the spacecraft would send down until the day after closest approach. This was the "fail-safe" data set designed to ensure that even if New Horizons was destroyed by some undetected debris, there would be a small but valuable scientific return from the mission.

After that, for more than a day, New Horizons would be far too busy taking data close to Pluto to look away and point its dish antenna to communicate again with Earth. Instead, it would be doing what it was built to do: photographing Pluto's surface in detail; mapping its surface composition; studying its atmosphere; then turning to image Pluto's giant moon, Charon, and briefly studying each of Pluto's four small moons. Some 236 separate scientific observations of each of the six bodies in the Pluto system, using all seven of the New Horizons instruments, were made over the next roughly thirty hours.

All the fail-safe data—including new composition spectra, atmospheric spectra, and this big Pluto image—reached Earth just before midnight on July 13. The crown jewel of the fail-safe data was a single full-frame, black-and-white image of one entire hemisphere of Pluto, taken just before the transmission began. By far, it would not be the most detailed image of Pluto that New Horizons would get, but it was made from a range three times closer than any previous image sent to Earth and was at the time the highest-resolution and most spectacular image of Pluto ever taken.

John Spencer led a handpicked team of just five New Horizons scientists, including Hal Weaver, who would process this image. They stayed up most of the night, getting it ready to release to the world the following morning. If there ever was a group of scientists who did not mind pulling an all-nighter, this group was it. John recalls:

> We already felt we were in a very privileged position just to be part of New Horizons at all, but to be the five people who were first seeing this planet that billions of people were waiting to see was amazing. And we couldn't believe what we were seeing in that image: you immediately could see that some parts of Pluto's surface were heavily cratered and ancient, and that other areas seemed remarkably crater-free, and therefore much younger. There was a huge range of ages across the surface, something almost unprecedented!

Alan:

> I wasn't able to be part of the overnight image processing team, because I would have to be ready for a twenty-four-hour-plus day of media coverage, press conferences, and mission critical operations the next day. I think my schedule allowed a four-hour window for sleep that night. The next morning, though, when I saw that jaw-dropping, fail-safe image of Pluto, I was simply stunned. This wasn't another fuzzy image made from too great a distance that only showed vague details—this one was razor sharp, and for the first time revealed Pluto's amazing geological beauty. With that image, Pluto became a place, just like Mars or Titan or even Earth, and it revealed itself to be a place beyond my wildest imagination. You could see mountain ranges, craters, canyons, giant ice fields, and more! It was amazing: Pluto was gorgeous. We'd hit the jackpot. I couldn't take my eyes off it.

16

EVEREST

A FINAL COUNTDOWN

The actual moment of closest approach to Pluto was to occur at 7:49:50 A.M. EST on Tuesday morning, July 14. But when this moment came—strangely for such an important event—there was nothing to see, and nothing new to show or to know. As we described before, the spacecraft was busy taking data, rather than communicating with Earth. It was skimming less than eight thousand miles above the surface of Pluto, working feverishly to get the goods it had come for. Back at APL, the New Horizons team marked the moment with a televised public countdown to celebrate the milestone.

The giant auditorium and the cavernous overflow area at APL's Kossiakoff Center were brimming beyond capacity, with bodies crammed in at the fire marshal's legal limit. Digital displays showed hours, minutes, and seconds left to reach Pluto, now all down to zeros except for the seconds counter. As those final seconds counted away, Alan led the crowd in a ten . . . nine . . . eight countdown rally. Each number was shouted in unison by the mission team and the massive swarm of space fans. At the second before closest flyby, a roar of cheers erupted, turning into a sea of smiles and furiously waving American flags.

At the same moment, at T–0, as New Horizons was passing closest approach to the planet it had traveled so far to explore, the full-frame fail-safe image of Pluto, which had been publicly released on the internet barely an hour earlier, was projected onto the jumbo display where the approach countdown had been, creating a visceral feeling in the room of being there, at Pluto, in the very moment.

Shivers ran up spines. Some people hooted or cheered. Some cried. Alan, with a few of the team members and Clyde Tombaugh's children held up a poster-size copy of that old 1991 U.S. postage stamp that read PLUTO NOT YET EXPLORED. But on the poster, "NOT YET" had been crossed out so that it now read PLUTO . . . EXPLORED. Images of that moment also went viral.

Meanwhile, out there on its own, deep in the Pluto system, New Horizons was doing what it had been designed and built to do a dozen years before. As it flashed past Pluto and its five moons that day, it was gathering a library of data so vast that it would take sixteen months to send it all back to Earth. *If* New Horizons survived, and *if* the flyby all worked. But had it?

It would be fourteen hours before the phone-home message would be sent and then would reach Earth, 14 hours before the team and the world would know. . . .

WAITING

While everyone waited for New Horizons to check in with the phone-home message, there was a public to be served. So NASA and the New Horizons project had arranged for a full day of programming around the flyby. John Spencer:

> The rest of that day was just a blur of media, mostly. I spent pretty much the whole day at the Kossiakoff Center talking to journalists and TV people, as did pretty much everyone else on the team.

Alan recalls, "The interest was relentless. Everywhere we went that day and the next we were mobbed by long lines of journalists and autograph seekers."

It was probably just as well to keep the team busy with press during encounter day so as not to spend their time worrying about their spacecraft.

That afternoon, NASA hosted a public panel discussion with science team members describing their first detailed impressions of Pluto and Charon based on the fail-safe images and other data. Jeff Moore discussed the overwhelmingly beautiful, aesthetic appearance of Pluto, pointing out that not even the most imaginative space artists had made a painting as stunning as the real thing.

And it was true. If you compare the actual Pluto we now know to artists' conceptions made before the flyby, none of the art comes anywhere near the striking splendor of the planet itself. In Jeff's words, it was another example of "Nature outdoing our imaginations every time."

Jeff then went on to describe some preliminary geological interpretations and showed a map with some of the new, informal names being tentatively used by the team to refer to features across Pluto's surface. The Kossiakoff audience applauded vigorously when he mentioned that the "whale" had been given a provisional name certain to please any fan of classic sci-fi: "Cthulhu."

Mission scientists Will Grundy of Lowell Observatory and Cathy Olkin of SwRI showed new color images. Will focused on the just-made discovery of Charon's dark northern polar cap—something completely unexpected, because it is unique in all the solar system. Will ventured a hypothesis that some of the escaping gases from the atmosphere of Pluto might be condensing out onto the cold north pole of Charon and then being chemically processed by solar ultraviolet light, producing organic molecules that mantled the pole of Charon. If this turned out to be correct, it would imply a startling physical bond between Pluto and its giant moon.

Cathy, a versatile researcher who had played many key roles on New Horizons, including as the deputy PI of the color imager inside the

Ralph instrument, showed a "stretched" (i.e., digitally exaggerated) color image of Pluto that demonstrated how different places on Pluto exhibited distinctly different colors. That image—which Cathy described as "psychedelic"—was greeted by gasps and a wave of wows from the audience. She then showed that Pluto's heart also has two very different colors, with the western lobe being whiter than the eastern lobe, which was markedly more blue. Also in that color-stretched view, Cathy noted that Pluto's north pole could be seen to be more yellow than the rest of the planet.

Randy Gladstone, also from SwRI and the head of the New Horizons atmospheres science theme team, then reported that Pluto's diameter had been measured to be 1,472 miles (later revised upward to 1,476 miles): larger than almost anyone had predicted. Alan then took pleasure in pointing out that this meant Pluto was, after all, larger than any other of the small planets in the Kuiper Belt, laying to rest the hope by some that dwarf planet Eris was larger than Pluto. As to those hoping Pluto was the second-largest body in the Kuiper Belt, Alan declared to the combined in-person and broadcast audience, "Well, now we can dispose of that."

Then the program was opened up to questions. Someone asked, "Why, in the largest crater seen on Charon, is the rim bright but the interior so dark?" John Spencer replied that the crater interior is the area that receives the most heat during an impact, and that there could be something about the material Charon is made of that turns dark when it is heated during crater forming impacts. Then he added, wryly, "That's one theory; but I just thought of it, which probably means it's wrong." That got laughs and applause, because it perfectly captured the atmosphere of instant, speculative science that occurs during a spacecraft flyby of a new planet, when the data is coming down faster than anyone's ability to make careful science of it.

OUTBOUND AND PREGNANT

After the afternoon science panel adjourned, most people went out for quick, early dinners and then returned to the Kossiakoff Center for the critical "phone home" spacecraft check-in moment later that evening. NASA televised the event, just like the flyby countdown that morning.

At Kossiakoff that evening, the press, the VIPs, the NASA brass, and guests all gathered to watch in the main auditorium. More than another thousand people were in the overflow areas. Giant screens showed the Mission Operations Manager Alice Bowman and her team in the MOC attending their consoles, waiting for the crucial signal to come down at the speed of light from Pluto, by way of NASA's Deep Space Network.

Alan, his assistant Cindy Conrad, Glen Fountain, NASA Planetary Exploration Director Jim Green, Green's boss John Grunsfeld, and NASA Administrator Charlie Bolden were just outside the MOC in the neighboring conference room, looking in through a glass wall and discussing the mission's significance.

A larger-than-life image of Alice was displayed on the Kossiakoff's jumbo monitor, scanning her array of computer consoles, waiting and listening intently to her headset. The moment approached when New Horizons would either make contact or leave a deafening, conspicuous silence. Then, within seconds of the scheduled time of 9:02 P.M., displays on the MOC's big board computer display began to show data about to flow in.

The big-board MOC display screen began to fill with numbers and telemetry messages, all colored green—no red anywhere! In the room adjacent to the MOC where he watched, Alan told NASA Administrator Bolden, "Look, look at that—New Horizons is healthy. The encounter succeeded!"

Alice in a clear, matter-of-fact voice, reported the unfolding events for the world:

"*Okay, we are in lock with carrier* [signal]. *Stand by for telemetry. In lock on symbols. . . . Okay, copy that. We are in lock with telemetry with the spacecraft.*"

Applause rang out among Alice's colleagues in the MOC. In the conference room next door, the television showed handshakes and high fives and hugs between Alan, the NASA Administrator, and others. New Horizons had survived the close approach to Pluto!

The applause quickly spread, along with loud, exuberant "woohoos" across the Kossiakoff auditorium and its overflow rooms.

Next, the voices of the various MOC engineering console operators were heard reporting to Alice, followed by her acknowledgment:

"MOM, this is RF on Pluto One."

"Go ahead RF."

"RF is reporting nominal carrier power, nominal signal-to-noise ratio for the telemetry. RF nominal."

"Copy that RF nominal."

"MOM, this is Autonomy on Pluto One."

"Go ahead, Autonomy."

"Autonomy is very happy to report nominal status. No rules have fired."

Here there was a smattering of applause, from those who understood its meaning: New Horizons had encountered no problems requiring it to undertake any kind of emergency response. Again, the applause spread outward to the crowds watching the projected video, who picked up on the engineer's glee at what must be very good news.

"C&DH."

"Go ahead, C&DH."

"C&DH reports nominal status, our SSR pointers are where we expect them to be, which means we recorded the expected amount of data."

"Copy that; looks like we have a good data report!"

As Alice said these last words, her voice swelled and she could be seen breaking into a wide smile. New Horizons was reporting that the amount of data in its solid-state recorder was just what it should be if it had made every one of its intended Pluto system observations. Still more applause: everything had worked!

Over the next minute or so, the rest of the subsystem teams came in with the remaining reports of nominal performance during the flyby. Alice concluded with a summary report to Alan on the communications loop: *"PI, this is MOM on Pluto One. We have a healthy spacecraft. We've recorded data at the Pluto system, and we're outbound from Pluto."*

At the very instant Alice finished speaking, the door flew open from the glass-walled conference room next door to the MOC. Alan breezed into the control room, grinning and beaming, arms raised high, pumping his fists. He went directly to Alice and they hugged.

The crowd in the MOC and in the Kossiakoff Center went wild with a sustained standing ovation. Alan whispered into Alice's ear something no one could hear: "We did it, we did it!" He fought back tears. "Flying across the solar system and exploring Pluto with you has been the honor of a lifetime."

Still on screen, Alan could be seen turning to shake hands and slap backs with Chris Hersman and others around the MOC as the elated applause continued. One of the announcers mumbled into a hot microphone, "Boy, I'm gonna lose it." Alice could be heard saying, "Sorry. I can't express how I feel. I'm shaking. Just like we planned it. Just like we practiced. I mean . . . we did it!" And then she giggled.

Minutes later, Alice, Alan, and all the others in the MOC left to travel across the APL campus to the Kossiakoff Center. On his way, Alan tweeted, "Don't know about you, but I had a pretty good day today. #PlutoFlyby"

As they arrived at Kossiakoff around 9:20 P.M., the MC asked the crowd to welcome the New Horizons team. Everyone there turned toward the top of the auditorium, craning their necks to see, and in came Alan, followed by the NASA administrator and John Grunsfeld, the head of all NASA science missions, followed in turn by Glen Fountain and then a long line with dozens of team members from the MOC, the engineering team, and the science team, all in their New Horizons mission shirts, walking down to the auditorium floor in single file.

As each team member entered and headed down the aisle, they high-fived bystanders and then Alan and Alice and Glen, who had gathered to welcome them to the floor at the front of the auditorium. The crowd responded with a standing ovation for three solid minutes as the team filed in: they had become space rock stars.

At the podium, NASA administrator Charlie Bolden announced, "We have now visited all the planets of the solar system!" And the audience responded with a "Pluto salute," hundreds of people holding nine fingers in the air.

At the same time, three billion miles away, New Horizons was outbound from Pluto, but still gathering important data. Its recorders were now pregnant with the prize so many had worked so long for—a scientific treasure trove of data that would revolutionize knowledge about Pluto, its moons, and the nature of all the small planets in the Kuiper Belt.

A BONFIRE IN MARYLAND

New Horizons was speeding away from Pluto, but its work was far from finished. Just about the time NASA Administrator Bolden was addressing the crowd at APL, New Horizons was turning to take images of Pluto's backlit orb to search for atmospheric hazes. There were many more science observations to come that night and in the coming days, but most important to the press and public was the knowledge that by morning there would be first-look images from the closest approach. And with that transmission and the new data it would bring, tomorrow would be another long workday for the New Horizons team, and it would begin startlingly early.

But no matter! The New Horizons team knew that their flyby was in the bag. So there in Maryland, it was their time to party. After escaping the crowds at APL, the team, with family and friends gathered back at the long-familiar nearby Sheraton, where most of the out-of-town team was staying, and headed directly to "Ten Forward," their private conference room and makeshift party suite.

The party was already well under way when Alan, who had been held up with media interviews at Kossiakoff, entered to a standing ovation. That recognition by his team, family, and colleagues was sweeter than any other he remembers.

At some point during the ensuing party, Alan and a dozen others ended up down by the hotel's swimming pool to reenact a scene from their Florida launch party nearly ten years before, in which the ULA rocket team had conducted a ritual bonfire burning of the launch malfunction procedures. Alan:

> I remembered that great, celebratory rocket-science ritual back at the launch party. So just before the flyby, I reminded some of the team about it and said, "If everything works, let's do that again: Let's go outside after we get back, and build a bonfire in a trash can out by the swimming pool and burn our plans for responding to flyby anomalies.
>
> So—fueled by a good amount of alcohol—we went down by the pool and built a fire, and threw those now-useless anomaly-procedure documents into it, laughing and savoring the moment.

"SOMETHING WONDERFUL"

The next morning, just as scheduled, the first of the truly high-resolution images of various locales on Pluto were received on Earth. Alan:

> Those first high-resolution images proved to be scientific gold, even beyond our expectations. I was bowled over by the complexity of the scenes—so much was going on in each and every patch of Pluto's surface. When I saw that, I remember thinking that everything we'd done to get there, all the career and personal sacrifices, were suddenly vindicated, the whole freaking twenty-six-plus years had been worth it.

And with those images, Alan knew they had something worthy of that *New York Times* cover page that they'd waited so long to grace, and, indeed, the next morning's *Times* featured New Horizons on the cover, above the fold with giant headlines. Almost five hundred other newspapers around the world did the same that day. The Pluto flyby was everywhere in the news.

As July 15 continued to unfold, another high-resolution image, equally stunning, and then another, and another, assembled on the science team's computers. One showed the surface of the western part of Pluto's vast heart—a region larger than the state of Texas. It showed an intricate and strangely organized geological pattern that left the normally loquacious New Horizons geologists at a loss for words. The scene contained smooth, bright areas, separated by narrow channels or ridges, which were vaguely polygonal, suggesting slow-motion convection cells, like the pattern seen on the surface of a heated, churning liquid. But how could that be, in this cold, cold place whose surface temperature was 400 degrees Fahrenheit below 0? Perhaps instead it was some sort of "polygonal cracking" like geologists see in ice-laden regions on Earth or Mars, where repeated freezing and thawing have led to regular patterns of cracks in mud or ice. Whatever it was, it appeared there was something amazing happening there. Something was moving and changing and flowing on the surface over time. Alan:

> I remember thinking, "This little planet is truly a spectacular place." It rivals or beats many of the larger planets in geological complexity. Before the flyby, I could not in my wildest dreams have pictured structures like these or imagined how strong Pluto's geological personality would turn out to be. It was just astounding.

At a NASA press conference broadcast later that day, a panel of New Horizons scientists spoke in front of another packed Kossiakoff auditorium and a massive online NASA TV and internet audience. Alan kicked things off with a tongue-in-cheek understatement that echoed his tweet the night before: "Well, I had a pretty good day yesterday. How about you?" Then he described how New Horizons was now already more than a million miles on the other side of Pluto, and that it was beginning to send us the first of the many treasures it would be returning to Earth over the next sixteen months.

Mission scientist Hal Weaver presented images showing surface details of Pluto's small, outermost moon, Hydra, revealing its size and shape for the first time. Those images showed that Hydra was elongated, and somewhat potato-shaped, but with axes twenty-eight by nineteen miles across. Hal then described their discovery that Hydra's reflectivity was very high, like freshly driven snow, suggesting that its surface is likely composed of water ice.

Next, New Horizons Composition science theme team leader Will Grundy reported on the first preliminary composition maps of Pluto, which showed strong variations in methane ice abundance across different geological regions. Will also reported that it was already clear that his team was seeing even more stunning diversity in composition, with different molecular ices—nitrogen, methane, and others—varying in abundance across different places on Pluto.

Deputy project scientist Cathy Olkin smiled as she showed off a stunningly beautiful new image of the close-approach hemisphere of Charon. "I thought Charon might only show ancient terrain covered in craters. Many on the team thought that might be the case. But Charon just blew our socks off when we got this new image today." She then led the audience through a brief tour of that newly revealed world, one "with deep canyons, troughs, cliffs, and dark regions that are still mysterious to us. We've been saying that Pluto did not disappoint. I can add that Charon did not disappoint us either."

Next, John Spencer made a carefully crafted announcement on behalf of the entire team: "We now have a name for Pluto's heart: We want it to honor the discoverer of Pluto, so we are calling it Tombaugh Regio." With that, the audience broke into applause, and the NASA camera cut to Annette and Alden Tombaugh, Clyde's retirement-age children, front and center in the audience. They were beaming. Then Alan added, "We could see Pluto's heart from very far away. When we were still 70 million miles out and only barely resolving the planet, we could see it shining like a beacon. Because it's the most prominent feature on the planet, we're going to name it in Clyde Tombaugh's honor."

Then it was time for John Spencer to reveal the pièce de résistance—the first very high-resolution images that science team members had gawked at when they first appeared on John's laptop a few hours before. It revealed the southwest corner of Tombaugh Regio, where Pluto's heart abuts the adjoining darker mountains of Cthulhu. The image showed steep and starkly shadowed mountains 3 billion miles from Earth. As the audience gasped and then cheered, John quipped, "That was our reaction, too!"

John then explained how the team could deduce the heights of mountains from the length of the shadows they cast. "These mountains here that we're seeing are spectacular . . . up to 11,000 feet high. They look to be tens of miles wide. So these are pretty substantial mountains. They stand up against the Rocky Mountains and other significant ranges here on Earth."

The implications of such high, sharp, and fresh-looking mountains for the nature of Pluto were profound. Scientists had long known there was a lot of nitrogen and methane on Pluto's surface, but that mountains could not be made out of those substances, because solid nitrogen and methane are simply not strong enough materials to support such steep relief, even in Pluto's low gravity. As a result, nitrogen or methane mountains would slump down under their own weight. No, these mountains had to be made out of something stronger—most likely water ice, which is the most common surface material on the satellites and other worlds of the outer solar system. These landforms implied that giant blocks of the water-ice "bedrock" from Pluto's crust had somehow been displaced and forced upward into dramatic mountain ranges. Next John showed a close-up of Tombaugh Regio.

> This scene is about 150 miles across. We see features as small as half a mile here, so you could see the APL campus on this image. The most striking thing geologically is that we haven't found an impact crater on this image. That means this is a young surface. Just eyeballing it, we think it has to be probably less than 100 million years old, which is a tiny fraction of the 4.5-billion-year age of the solar system. I never would have believed that the first close-up picture we got of Pluto would not have a single impact crater on it. This is astonishing.

The big question was: What is causing all the geological and tectonic activity they'd discovered? Why the lack of craters in some areas, the wide variety of textures and compositions, and the huge mountain ranges?

They were all telling the same story—that while Pluto itself is ancient, its surface is young and active. New Horizons was revealing that Pluto is capable of supporting active geology more than 4 billion years after its formation. But how? Textbook geophysical theory predicted that a small planet like Pluto should have long ago cooled off and ceased making new surface geology. But the data was irrefutable. Pluto, it seems, hadn't read the textbooks.

As the press conference neared its end, Chip Reid from CBS News directed a question at Alan: "I interviewed you years before the flyby, and the only prediction you would make is that we would see something wonderful. Have your expectations been met?"

With a smirk, Alan responded, "I'll give you a technical answer: *Ya think?*"

GOING VIRAL

New Horizons attracted an unusually high level of media and public attention before the flyby. But when the world finally saw the stunning quality of the images and the beautiful and photogenic nature of the planet they revealed—with its dramatic topography, strange surfaces, and, of course, its bright heart—that attention multiplied to levels NASA had never seen. The worldwide reaction in the wake of the flyby was instantaneous and simply unprecedented.

On the morning of July 16, fifty years to the very day after the first pictures from the first flyby of Mars dominated the front page of *The New York Times* in 1965, the paper ran a huge New Horizons image, on the front page above the fold. Giant images of Pluto were projected to crowds at Times Square. The internet went wild for New Horizons as well, with Google even doing a special animated Pluto "doodle" on their home page, in which the second *O* was replaced with a stylized spinning Pluto (embossed, naturally, with a heart) and a little cartoon of New Horizons arcing across the frame.

NASA had experienced large online reactions before, logging more than 100 million hits for Mars landings, but it had never seen any-thing like this. The day of the flyby was NASA's biggest day ever on social media and websites, receiving over 1 billion hits. New Horizons rocked Facebook and Twitter and trended at the top of Instagram for days. Scores of memes, many riffing on the heart motif, were created and bounced around the internet. The soaring media reaction even spurred its own metareaction, with dozens of stories written about the level of attention that had been generated.

The ubiquity of New Horizons and Pluto on the web, and the num-ber of people sharing in New Horizons events around the globe, gave this flyby an entirely new kind of feel. The world had changed since Voyager, with so many new forms of communication and participa-tion. Thanks to that, the New Horizons mission felt in many ways like the first truly twenty-first-century planetary encounter.

Consider: with Voyager, to participate fully you had to be in just the right place—specifically, at JPL—at just the right time—on flyby day. For New Horizons you didn't need to be there; the flyby was everywhere simultaneously. The events at APL, the imagery from Pluto—everything that reached Earth—went onto the internet "for all mankind," as it were.

Yes, it was undeniably cool for those able to be present at APL, the crowd of enthusiastic space nerds, the mission team, the press circus, the politicians, and the celebrities. But even the people at APL were themselves actually spending a lot of time online, looking at and partici-pating via the internet and social media, sharing images, information, and impressions in real time with a worldwide community that was seeing it all, and chiming in, as it happened. Even though the flyby was taking place 3 billion miles away, it felt in some ways as though people were able to cheat both the speed of light and the expansive scope of the Earth. It felt as if the global mass of humanity was all there together.

The word "together" here is important. It's not just that everyone was receiving a broadcast at once. The sense of participation and com-munication, with people from all over contributing to the conversa-tion via social media was very real, and unlike anything seen in the

first flybys of all the closer planets back in the twentieth century. With
New Horizons, humankind was able to directly share the flyby and
the very human events around it, just as it all happened, transforming
it into a communal, worldwide experience.

EVEREST, SUMMITED

Following a theme of honoring the explorers who came before them,
the New Horizons team gave two ranges of soaring water-ice mountains along the western side of Tombaugh Regio the names Norgay
and Hillary Montes, after Tenzing Norgay and Edmund Hillary, the
first explorers to reach the top of Everest, Earth's highest peak.

This association between those who summited Everest and those
who summited Pluto was particularly apt. Alan had been referring to
Pluto as the "Everest of the Solar System" since the 1990s—meaning
that it was the last, the farthest, the coldest, and the hardest peak of
planetary exploration.

But what Alan didn't anticipate, or realize until the actual flyby and
its immediate aftermath, was how, when the moment came, he would
feel like he imagined those earlier explorers must have felt upon sum-
miting Everest itself. Alan:

> When I think back to the few days following the flyby, it really
> felt like we'd all had a peak experience in our lives. We'd
> reached and then summited our own metaphorical mountain,
> Pluto.
>
> And you know, more than anything, it took an amazing team of
> people that worked together for a very long time to achieve this—
> something vastly larger than any of us could have accomplished
> individually. Across our team during the flyby there was very
> much a feeling of being part of a band of humans out exploring,
> accomplishing something extraordinarily special. So many of us
> said to one another that week what a privilege it had been to
> help make the flyby happen and to be able to inspire others per-
> haps to one day do even greater things in space exploration.

Twenty-six years had elapsed between that first fateful meeting to discuss the idea of going to Pluto with NASA in May of 1989, and that summer in 2015 when the exploration of Pluto was accomplished. People who were not even born when it started were moved by it in ways that no one had imagined when the quest began.

History was made. New knowledge was created. A nation was reminded it can achieve greatness. And a world was reminded that we humans, we Earthlings—really can accomplish amazing things.

17

ONWARD NEW HORIZONS

LOOKING BACK FROM THE FAR SIDE OF PLUTO

For most humans, vision is the most impactful of all our senses, and of all the valuable data returned by spacecraft, it is the images that move us the most. Among all the mind-blowing images of the Pluto system taken by New Horizons, we of course have our own favorites. Many are color images made by the Ralph instrument, but others are black-and-white LORRI images. One is the gorgeous high-resolution color face of Pluto and its vast heart-shaped Tombaugh Regio that graces the cover of this book. Another is the montage of Pluto and Charon as a double-planet system. We also love the color images of methane-snowcapped water-ice mountain ranges.

Our favorite black-and-white image of the flyby is of a high-resolution crescent Pluto, taken just fifteen minutes after New Horizons made its closest approach (see the photo insert). Why? Because it so vividly shows Pluto as a world with dramatic topography and concentric haze layers stretching half a million feet into its sky. That vista also reveals dramatic flows swirling across the surface of the gigantic, nitrogen-ice glacier named Sputnik Planitia, bordered by the towering mountain ranges of Norgay and Hillary Montes, their long shadows highlighting the planet's rugged topography. Shot in sunlight a thousand times dimmer than the Sun shines here on Earth, the image is riveting both for capturing the alien beauty of the world called Pluto, as well as for what its existence captures about humans and our drive to explore. As New Horizons team scientist Cathy Olkin said, "This image makes you really feel like you're there."

Our favorite color image from the flyby is completely different, though it was also taken shortly after closest approach. That one, made about an hour after the crescent image just mentioned, was taken as New Horizons flew through Pluto's shadow during the occultation experiments used to probe the planet's atmosphere. This stunning shot (again, see the photo insert) reveals that Pluto's atmosphere, perfused with sunlight, is a deep blue, just like Earth's.

But we love this particular image most for another reason beyond its simple beauty. When Apollo astronauts first orbited the moon in 1968, they took pictures of the Earth rising above the Moon's limb, and it defined the achievement of humans traveling off our planet, to the Moon, and it made humanity appreciate our own planet anew and marvel at what we humans can accomplish.

For us the blue backlit Pluto image evokes the same emotions that Apollo's Earthrise image did. And we love the symmetry of the front-lit Apollo Earthrise image taken at the dawn of the era of planetary exploration, when most of the planets were still unexplored, complemented by the backlit far-side image of Pluto, taken on the very day that the capstone was placed on the first era of planetary reconnaissance—July 14, 2015.

When we look at this sublime image, we also think about how it was made and what it represents: that's Pluto, backlit by the Sun. Just like the Apollo Earthrise image could only be made from the vicinity of the Moon, this image could only be made from the far side of Pluto.

As described in this book, the exploration of Pluto by New Horizons could have failed for so many reasons. It so easily could have failed to get funded. And New Horizons itself, by all rights, should not have been selected given its underdog, less experienced team going up against more experienced competitors. Given the short development time allotted and a budget only one-fifth of Voyager's, New Horizons probably could also have failed to get built and then launched on budget and on time, or even at all. And it might have failed along the journey and not gotten the goods at Pluto. But it didn't fail. Instead it succeeded brilliantly—thanks to an amazing team, to determination, ingenuity, pluck, persistence, and some lucky breaks.

The people who created this amazing mission of exploration chased their new horizons hard; they never let go of their dream; they put everything they had into it; and eventually they chased it down and accomplished what they set out to do. To us the bluish image taken looking back at Pluto backlit by the Sun symbolizes the accomplishment of the exploration of Pluto.

Look at it again. We did it. We really did. We were there.

LOOKING AHEAD

After leaving the Pluto system, New Horizons was approved by NASA for a five-year, extended mission to study other bodies in the Kuiper Belt. The centerpiece of that mission will be the flyby of an ancient Kuiper Belt Object (KBO) that represents the building blocks of small planets like Pluto. (The flyby's target is one of the objects found by using the Hubble, in a dramatic search for flyby targets after Pluto, described in chapter 13).

That next flyby will take place on New Year's Eve and New Year's Day 2019, a billion miles farther out than Pluto. The target, called 2014 MU69, is only about twenty miles across, but it appears to be a binary, just as Pluto and Charon are. New Horizons is planning to fly by MU69 at a range of barely two thousand miles, almost four times closer than its flyby of Pluto.

In addition to mapping MU69, studying its composition, and searching for moons and any atmosphere, as it travels outward through the Kuiper Belt, New Horizons will also study another two dozen or more KBOs from afar using its LORRI telescope/imager. Those studies will be used to search for satellites and rings, determine surface properties, rotation periods, and shapes for those objects, in order to put the close-up study of MU69 in better context. During this extended mission, New Horizons will also be used to conduct a five-year-long study of the environment in the Kuiper Belt by continuously monitoring the charged particles and dust there as it crosses out to a distance of fifty times as far from the Sun as Earth—the far limit of Pluto's orbit. New Horizons will reach that distance in April of 2021, at which time it is scheduled to be turned off by radio command from Earth.

However, fuel and power projections indicate that New Horizons will be able to continue to operate in the service of exploration into the mid-2030s or even later. (That may be longer, we note, than many of its own builders operate.) So if NASA continues to fund the mission, New Horizons will follow the legacies of Pioneer 10 and 11 and Voyager 1 and 2—becoming a probe of the Sun's distant heliosphere and the nearest reaches of interstellar space.

Then someday, perhaps in the late 2030s, or perhaps in the 2040s, there will not be enough power to run the spacecraft's main computer and communications system, and New Horizons will fall silent. Nonetheless, it will continue its journey, speeding forever away from the Sun into interstellar space: a derelict—yes—but also an immortal denizen of the galaxy, and an emblem of what humans can achieve.

And what about Pluto? Will there someday be a return there and further exploration of that wonderland world and its fascinating moons?

We think so. There's a growing scientific consensus that the mysteries New Horizons revealed and the scientific questions it has raised cannot be fully answered until an orbiter, which can explore Pluto in much more detail, and perhaps even a lander, are sent there. Studies to explore how that can be accomplished are already under way, and the next Planetary Decadal Survey in the early 2020s is likely to consider such a mission. We're optimistic that a return to study the Pluto system in more depth will one day be funded. In addition, we think it's likely that the other small planets of the Kuiper Belt will also likely be explored by spacecraft later in this century.

If we humans are nothing else, we are an inquisitive and restless species, explorers at heart. For that reason, we're also optimistic that even humans will one day travel to the Kuiper Belt to explore it in person, making footfall on Pluto and other Kuiper Belt worlds, as we have already done on the Moon and will soon do on Mars, and then no doubt on many other worlds.

The first exploration of Pluto is complete, but the call of exploration beckons our species ever onward, into the wild black yonder of our solar system.

CODA

A FINAL DISCOVERY

The heroes of New Horizons are the engineers, scientists, and others who worked so hard for so long to set and achieve a lofty goal, to discover new things about the wonderful universe we live in, to inspire, and to make their own contribution to what is called "history."

In accomplishing the exploration of Pluto, the New Horizons team set records and achieved many firsts. But more importantly, we think, they demonstrated to the world some of what are the best qualities of humankind: inquisitiveness, drive, persistence, and the ability to work in teams to achieve something larger than life.

Of those, more than anything else, New Horizons and the exploration of Pluto took persistence. Consider: it took thirteen years, countless battles, and six failed mission concepts just to win the funding to *start* building it. After that, it took another four years, galloping against all odds, to build and launch an outer-planets spacecraft in record time and at a breakthrough low cost. That in turn was followed by a marathon, nine-and-a-half-year journey across our entire solar system, by a lone robot and a small flight team on Earth, just to reach Pluto.

By exploring Pluto, New Horizons became the capstone mission to the initial reconnaissance of our vast, home solar system. In doing so, it turned the last of the planets known at the birth of the Space Age from a faraway point of light into a real place that humans have now come to know. And with that reveal, NASA, the United States, and our species completed a fifty-year-long quest to reconnoiter all the nine originally known planets—the space-age equivalent of Magellan's first circumnavigation of our home planet.

The exploration of Pluto was a scientific success beyond what almost anyone expected. It produced countless discoveries and upended paradigms, teaching us that small planets like Pluto can be as complex as big ones, and that small planets can remain intensely geologically alive even billions of years after forming.

The public reaction to the exploration of Pluto helped to reawaken something partially forgotten since Voyager and Apollo: that people across the world love bold space exploration, are inspired by missions to never before explored places, and that such missions even have the power to inspire people and change lives.

Shortly after New Horizons flew past Pluto, Alan gave a talk in Vermont. After he spoke, a college student told him that for too long her generation had been saddled with the meme that their time was not as great as those of past generations. She said that their generation hadn't witnessed wars that saved a world from fascism, that they missed the historic first steps on the Moon, the birth of computing, and so many other epochal events. Then she said that seeing Pluto explored was "our Moon landing, and the greatest thing that's happened in our generation." A shiver ran up Alan's spine when she said that, and he realized that in her eyes the New Horizons mission had been successful in a way he had never before imagined.

A few months later, after Alan gave a talk to a business convention in Florida, he was approached by a middle-aged woman who came to him literally in tears. She explained that her teenage son had been a failing student until he saw the flyby of Pluto by New Horizons, and that Pluto's exploration had inspired him to say, "That's what I want to do when I grow up." The mother wiped a tear and told Alan that her son had since transformed himself into a straight-A student. She said, "You all rescued my son."

We believe that the power of these and other very human impacts made by New Horizons outshine everything learned about Pluto. And for us, nothing can substitute for that discovery.

—DAVID GRINSPOON, Washington, DC

—ALAN STERN, Boulder, CO

APPENDIX

THE TOP TEN SCIENCE DISCOVERIES FROM THE NEW HORIZONS EXPLORATION OF THE PLUTO SYSTEM

Research papers describing and analyzing scientific results from the New Horizons flyby of Pluto, Charon, and Pluto's other moons already fill volumes in peer-reviewed journals of planetary science. And there is sure to be much more scientific yield to come over the years, as understanding of the results grows and is integrated into our overall understanding of the origin and evolution of planets. What follows here (not in any ranked order) is a brief recapitulation of the ten most important discoveries made by the New Horizons mission, as named by NASA and the New Horizons team in 2016. For each of these "top ten" discoveries, we provide a little elaboration.

THE SHEER COMPLEXITY OF PLUTO

The diversity of phenomena seen on Pluto was far beyond what anyone, even New Horizons team members, expected to find on such a small planet so cold and far from the Sun. Ground fogs, high-altitude hazes, possible clouds, canyons, towering mountains, faults, polar caps, apparent dune fields, suspected ice volcanoes, glaciers, evidence for flowing (and even standing) liquids in the past, and more. This little red planet perched 3 billion miles away in the Kuiper Belt packed more punch than any other known small world explored, and indeed more punch than many much larger worlds. The variety of terrains, its complex interactions between the surface and the atmosphere, and the wide range of surface ages even prompted the New Horizons team to adopt the slogan "Pluto is the new Mars."

THE STUNNING DEGREE OF LONG-TERM AND CONTINUING ACTIVITY ACROSS PLUTO'S SURFACE

There were many reasons that some expected Pluto to be a relatively dead world, geologically speaking. After all, it is such a small planet, and it lacks the tidal heat sources that being in a giant planet's satellite system would provide. Also, it is far from the Sun and solar heating is weak. So, by the conventional wisdom born of exploring the rest of the solar system, Pluto should have been largely or even completely geologically inactive for eons. But the conventional wisdom was seriously wrong. New Horizons found a wide range of surface ages, ranging from ancient and heavily cratered to completely fresh-looking areas with no craters at all—meaning that Pluto has been geologically active throughout its 4-billion-year history. In fact, Pluto has been alive and kicking throughout history, and is even today. Why that is so is the subject of intense scientific debate and modeling, and it portends that we can expect more surprises when other small planets in the Kuiper Belt are explored.

THE VAST, 1,000-KILOMETER-WIDE SPUTNIK PLANITIA NITROGEN GLACIER

Among the varied and active terrains discovered on Pluto, probably the most remarkable of all is the vast ice field of Sputnik Planitia, which revealed itself to be actively convecting, churning like a slow-motion pan of sauce cooking on a stove. Sputnik Planitia is now understood to be a vast and deep layer of nitrogen ice, spiked with methane and carbon monoxide, resting within a huge bowl-shaped depression, which is most likely an ancient impact basin. Nitrogen glaciers feed into it from the mountainous areas around its periphery. Giant water-ice icebergs float along its margins. In a way, Sputnik Planitia is like a frozen ocean of nitrogen on Pluto's surface. Sputnik Planitia—the western lobe of Pluto's heart—dominates every view of the flyby hemisphere. But it also seems to play a deeper role within the strangely coupled geology and meteorology of Pluto. It is a huge reservoir of nitrogen, the primary substance that is transported through the atmosphere and onto and across Pluto's icy, cold surface. Depending on the season and climate phase that Pluto is in, the amount of nitrogen fluctuates dramatically between atmosphere and surface. So Sputnik Planitia may at times be deeply eroded and at other times be recharged with even more nitrogen—and there is evidence for this in the currently empty glacial channels in the surrounding highlands. Nothing like Sputnik Planitia exists anywhere else in the solar system.

THE DISCOVERY OF EXTENSIVE, WELL-ORGANIZED ATMOSPHERIC HAZES

The spectacular "look back" images of Pluto, illuminated by the Sun from behind the planet as New Horizons was racing away after the flyby, vividly show Pluto's beautiful blue atmosphere. But these dramatic images also reveal dozens of delicate layers of atmospheric haze hanging in Pluto's cold nitrogen air. These extensive hazes extend up at least 300 miles above Pluto's surface and are organized in concentric layers that stretch across many hundreds of miles of varied surface terrains. These hazes result from complex chemistry in Pluto's atmosphere, somewhat like the organic hazes seen in the atmosphere of Saturn's moon Titan. As on Titan, the methane in the air reacts with sunlight to make complex organic molecules which eventually rain out onto Pluto's surface, giving it its ruddy color.

A DRAMATICALLY LOWER ATMOSPHERIC ESCAPE RATE COMPARED TO EXPECTATIONS

All planets with atmospheres are constantly losing some amount of gas into space. Because Pluto is such a small planet, with low gravity and a low escape velocity, it was expected to be losing its methane and nitrogen atmosphere at a high rate. But one unanticipated and highly surprising result of New Horizons' measurements is that the loss rate of nitrogen was much slower than models had predicted— in fact, more than ten thousand times slower! The reason seems to be that the upper atmosphere is much colder than was expected. This means that the molecules are moving with lower velocities, so fewer of them are traveling fast enough to escape. The reason why the atmosphere is so cold is a mystery that has not been solved.

EVIDENCE FOR STRONG CHANGES IN ATMOSPHERIC PRESSURE AND THE PAST PRESENCE OF RUNNING OR STANDING LIQUID VOLATILES ON PLUTO'S SURFACE

We know that Pluto's atmospheric pressure varies exponentially with surface temperature. We also know that its atmospheric pressure should also change dramatically over time as it goes through climate cycles of millions of years and its orbit and spin slowly wobble, changing the angles and amounts of sunlight heating different parts of its surface. This was expected before the flyby, but New Horizons found several kinds of forensic evidence for much higher surface pressures in the past. These include areas of "washboard terrain," dunes, channels that might have been cut by flowing liquids, and even one distinct feature that looks like a frozen lake suspended in a mountain valley.

EVIDENCE FOR A POSSIBLE LIQUID WATER OCEAN INSIDE PLUTO TODAY

The giant glacier on Pluto called Sputnik Planitia is located almost exactly at the "anti-Charon point." That is, it is centered almost exactly opposite from the location on Pluto where Charon, being tidally locked, always hangs directly overhead. Why should it be precisely there? It's thought that the added weight of the ice in the Sputnik Planitia basin itself might have led to tidal forces that caused the basin to migrate into that position. However, that can only occur if the interior of Pluto and the planet's icy crust are "frictionally decoupled" by the existence of a liquid water ocean layer below the crust. Definitive tests for an interior ocean will have to wait for a future orbiter mission to Pluto, but even now we can ask, might this water ocean be inhabited? Could there be Plutonian life forms swimming deep underneath the planet's icy surface? Current thought within astrobiology holds that liquid water may be a key necessity for life, with organic molecules and some form of energy flow also required. It seems possible that all these con-ditions could be met within Pluto, as it is within other worlds with interior oceans, like Europa and Enceladus.

CHARON'S ENORMOUS EQUATORIAL TECTONIC BELT, HINTING AT AN ANCIENT INTERIOR OCEAN

Separating the northern from the southern hemispheres of Charon, and slanting at a sharp angle across the equator from the southwest to the northeast, is a huge complex of valleys and cliffs that stretch for over one thousand miles. Geological analysis reveals that this is a huge extensional belt, meaning that the surface of Charon was pulled apart across this yawning series of chasms by expansion forces. What would have caused Charon to split down the middle like that? In or-der to generate the forces sufficient to break apart its solid surface, it appears that Charon would have had to expand on the inside, like a beverage can that bursts when you leave it in the freezer for too long. What actually happened is probably similar to that. Charon's interior is about half water ice by mass, and we know that when Charon formed, its interior was hot, causing the water ice to be liquefied. Over time, as it cooled, the freezing process caused internal expansion, which in turn likely resulted in the creation of the widespread tectonic belt we see.

CHARON'S COMPLETELY UNIQUE, DARK, RED POLAR CAP

Perhaps Charon's most remarkable surface feature is its dark, reddish polar cap, also sometimes called a polar "stain," as it appears to be diffusely spread—almost sprayed on top of its underlying polar geology. Nothing like it has been seen anywhere else among all the other worlds of our solar system. The leading idea to explain the red cap is that some of the methane escaping off the top of Pluto's atmosphere strikes Charon and preferentially condenses on Charon's poles, the coldest places on this moon. As has been simulated in scientific laboratories on Earth, on Charon the methane can be processed by sunlight and solar wind into heavier hydrocarbon molecules which are dark and red—just as is observed on Charon's poles—and which are nonvolatile, meaning they don't evaporate. This strange connection between Pluto and Charon, with material from Pluto slowly migrating to Charon over the eons, is reminiscent of some kinds of binary stars that exchange material with one another gravitationally, and it adds to the oft-stated description of the Pluto system as a sci-fi lover's dream!

SMALL MOON MYSTERIES

Every aspect of the Pluto system contains its share of surprises. This has even been true of Pluto's four small moons—Nix, Hydra, Styx, and Kerberos—which orbit outside the Pluto-Charon binary. One surprising discovery about these small moons is how fast they are all spinning, much faster than their orbital periods. Hydra, the most extreme, rotates in just ten hours—almost one hundred times faster than its orbital period. Even stranger, their spin axes are not basically perpendicular to their orbital planes around Pluto, the norm in other systems of moons. Why? The answer is unknown. All four of these small satellites have the elongated, non-spherical shapes that are typical of icy objects not large and massive enough to pull themselves into a sphere with their own gravity. But yet another surprise is that two of them—Styx and Hydra—are each composed of two "lobes" that appear smashed together, which may be telling us that they formed by collisions of former moons. And if that's not enough, all four of Pluto's small moons are also surprisingly bright and reflective: they each reflect about 70–80 percent of the light that strikes them, which makes them among the most reflective objects in the solar system. And a final mystery is that despite an intense search for smaller moons by New Horizons, none were found. Given that almost every time the Hubble Space Telescope looked for new moons around Pluto, it found them, it surprised almost everyone on the New Horizons team that they didn't find more when the spacecraft was close enough to beat Hubble's capabilities. Why does Pluto have five moons and yet no more? No one knows.

ACKNOWLEDGMENTS

———

First and foremost, we wish to acknowledge the New Horizons team members, and the current and former NASA officials, and others, who agreed to be interviewed and quoted. Each generously lent us their time and insights; without their contributions this book would have not been possible. For this we particularly thank Fran Bagenal, Alice Bowman, Marc Buie, Glen Fountain, Dan Goldin, Mike Griffin, Chris Hersman, Wes Huntress, Tom Krimigis, Todd May, Bill McKinnon, Ralph McNutt, Jeff Moore, Cathy Olkin, John Spencer, Rob Staehle, Hal Weaver, and Leslie Young. We are also extremely grateful to all of the contributors to the success of New Horizons and the exploration of Pluto.

For other helpful conversations and correspondence, we thank Jim Bell, Laurie Cantillo, Candy Hansen, Charles Kolhase, Jonathan Lunine, Kelsi Singer, Joel and Leonard Stern, Chuck Tatro, the Tombaugh family, Stacy Weinstein, and Amanda Zangari. We are also indebted to Cindy Conrad, who was massively helpful to us, both logistically and editorially. And we also thank Morgaine McKibben, Michael Soluri, and Henry Throop for photographically capturing aspects of the New Horizons mission and allowing us to use their beautiful and oftentimes touching images. Thanks, too, to Kevin Schindler, Historian at Lowell Observatory, who generously shared with us archival materials that enriched this book, and Mike Buckley of the Applied Physics Laboratory, who gave invaluable help tracking down and providing NASA and APL images and other material. We also want to thank our agents, Carrie Hannigan, Josh Getzler, and Eric Lupfer, for skillfully guiding us through all the stages of this project, and to our editor, James Meader, for his constant encouragement, patience, and judgment, and for joining us in our excitement at sharing the story of New Horizons with the world. Finally, we thank our wives, Jennifer Goldsmith-Grinspoon and Carole Stern, for putting up with the many long weekends and evening hours that went into writing this book.

INDEX

ABOUT THE AUTHORS

DR. ALAN STERN is principal investigator (PI) of the New Horizons mission, leading NASA's exploration of the Pluto system and the Kuiper Belt. A planetary scientist, space program executive, aerospace consultant, and author, he has participated in over two dozen scientific space missions and has been involved at the highest levels in several aspects of American space exploration. Dr. Stern is the recipient of numerous awards, including the 2016 Carl Sagan Memorial Award of the American Astronautical Society, and has twice been named to the *Time* 100. He lives in Colorado.

DR. DAVID GRINSPOON is an astrobiologist, award-winning science communicator, and prize-winning author. In 2013 he was appointed the inaugural chair of astrobiology at the Library of Congress. He is a frequent advisor to NASA on space-exploration strategy and is on the science teams for several interplanetary spacecraft missions. Grinspoon's previous books include *Earth in Human Hands* (2016) and his writing has appeared in *The New York Times*, *Slate*, *Scientific American*, *Los Angeles Times*, and others. He lives in Washington, DC.

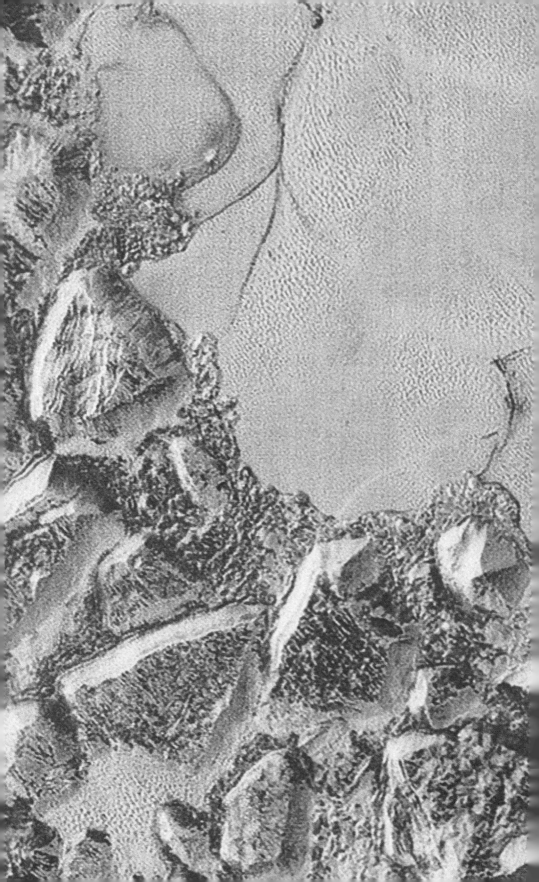